T0140329

Studies in Systems, Decision and Control

Volume 99

Series editor

Janusz Kacprzyk, Polish Academy of Sciences, Warsaw, Poland
e-mail: kacprzyk@ibspan.waw.pl

About this Series

The series "Studies in Systems, Decision and Control" (SSDC) covers both new developments and advances, as well as the state of the art, in the various areas of broadly perceived systems, decision making and control- quickly, up to date and with a high quality. The intent is to cover the theory, applications, and perspectives on the state of the art and future developments relevant to systems, decision making, control, complex processes and related areas, as embedded in the fields of engineering, computer science, physics, economics, social and life sciences, as well as the paradigms and methodologies behind them. The series contains monographs, textbooks, lecture notes and edited volumes in systems, decision making and control spanning the areas of Cyber-Physical Systems, Autonomous Systems, Sensor Networks, Control Systems, Energy Systems, Automotive Systems, Biological Systems, Vehicular Networking and Connected Vehicles, Aerospace Systems, Automation, Manufacturing, Smart Grids, Nonlinear Systems, Power Systems, Robotics, Social Systems, Economic Systems and other. Of particular value to both the contributors and the readership are the short publication timeframe and the world-wide distribution and exposure which enable both a wide and rapid dissemination of research output.

More information about this series at http://www.springer.com/series/13304

Joe Lorkowski · Vladik Kreinovich

Bounded Rationality in Decision Making Under Uncertainty: Towards Optimal Granularity

 Springer

Joe Lorkowski
Department of Computer Science, College
of Engineering
University of Texas at El Paso
El Paso, TX
USA

Vladik Kreinovich
Department of Computer Science, College
of Engineering
University of Texas at El Paso
El Paso, TX
USA

ISSN 2198-4182   ISSN 2198-4190   (electronic)
Studies in Systems, Decision and Control
ISBN 978-3-319-87260-5   ISBN 978-3-319-62214-9   (eBook)
DOI 10.1007/978-3-319-62214-9

Printed on acid-free paper

This Springer imprint is published by Springer Nature
The registered company is Springer International Publishing AG
The registered company address is: Gewerbestrasse 11, 6330 Cham, Switzerland

Contents

Abstract

Starting from the well-known studies by Kahmenan and Tversky, researchers have found many examples when our decision making seems to be irrational. We show that this seemingly irrational decision making can be explained if we take into account that human abilities to process information are limited. As a result, instead of the exact *values* of different quantities, we operate with *granules* that contain these values. Of several examples, we show that optimization under such granularity restriction indeed leads to observed human decision making. Thus, granularity helps explain seemingly irrational human decision making.

Similar arguments can be used to *explain* the success of *heuristic techniques* in expert decision making. We use these explanations to *predict* the *quality* of the resulting *decisions*. Finally, we explain how we can *improve* on the existing *heuristic techniques* by formulating and solving the corresponding optimization problems.

Chapter 1
Human Decisions Are Often Suboptimal: Phenomenon of Bounded Rationality

1.1 Seemingly Irrational Human Decision Making: Formulation of the Problem

Decisions are important. One of the main objectives of science and engineering is to help people make decisions.

For example, we try to predict weather, so that people will be able to dress properly (and take an umbrella if needed), and so that if a hurricane is coming, people can evacuate. We analyze quantum effects in semi-conductors so that engineers can design better computer chips. We analyze diseases so that medical doctors can select the best treatment, etc.

In complex situations, people need help in making their decisions. In simple situations, an average person can easily make a decision. For example, if the weather forecast predicts rain, one should take an umbrella, otherwise one should not.

In more complex situations, however, even when we know all the possible consequences of each action, it is not easy to make a decision. For example, in medicine, many treatments come with side effects: a surgery can sometimes result in a patient's death, immune system suppression can result in a infectious disease, etc. In such situations, it is not easy to compare different actions, and even skilled experts would appreciate computer-based help.

To help people make decisions, we need to analyze how people make decisions. One of the difficulties in designing computer-based systems which would help people make decisions is that to make such systems successful, we need to know what people want when they make decisions. Often, people cannot explain in precise terms why they have selected this or that alternative.

In such situations, we need to analyze how people actually make decisions, and then try to come up with formal descriptions which fit the observed behavior.

© Springer International Publishing AG 2018
J. Lorkowski and V. Kreinovich, *Bounded Rationality in Decision Making Under Uncertainty: Towards Optimal Granularity*, Studies in Systems, Decision and Control 99, DOI 10.1007/978-3-319-62214-9_1

In the ideal world, people should make perfect decisions. In many real-life situations, we know what is best for us, and we know the exact consequences of each of our actions. In this case, a rational person should select an action that leads to the best possible outcome.

This assumption underlies basic (idealized) economic models: in these models, our decision making may hurt others but every person is interested in selecting a decision which is the best for him/herself.

In the real world, people's decisions are not perfect. In the perfect world, people should make perfect decisions. It is well known, however, that our world is not perfect, and that many people make decisions which are not in their own best interests. People eat unhealthy food, fail to exercise, get drunk, smoke, take drugs, gamble, and do many other things which—as they perfectly know—are bad for their health and bad for their wallets.

Such imperfect decisions can still be described in optimization terms. People engage in all kinds of unhealthy and asocial decision making because they get a lot of positive emotions from this engagement. A drug addict may lose his money, his family, his job, his health—but he gets so much pleasure from his drugs that he cannot stop. A gambler may lose all his money, but the pleasure of gambling is so high that he continues gambling (and losing) money until no money is left.

These examples of bad decisions are bad from the viewpoint of a person's health or wealth or social status. In all these examples, people clearly know what they want—e.g., more pleasure from drugs or from gambling—and they select a decision which is the "best" from this viewpoint.

On top of this well-known abnormal decision making, there are also many examples when a seemingly rational decision is actually irrational. It is well known that people make seemingly irrational decisions as described above, i.e., that they are optimizing objective functions which lead to their physical and ethical ruin.

Somewhat less known for the general public—but well known in psychology—is the fact that many quite rational people, people who are not addicted to drugs or gambling, people who normally lead a reasonably rational life, often make decisions which, at first glance, may seem reasonable but which, on deeper analysis, are irrational. This was first discovered in the studies of the Nobel Prize Winner Daniel Kahneman and his coauthor Amos Tversky; see, e.g., [61].

What they discovered is that sometimes people behave in such a way that no optimization can explain. Let us give a simple example. When a person is given two alternatives A and B, the person is usually able to conclude that, e.g., to him or her, A is better than B. We may disagree—as is the case of drug addiction—that A is an unhealthy choice, but A is what a person prefers. In this case, if now we offer the same person yet another alternative C, he or she may stick to A, or he or she may switch to C—but we do not expect this person to select B (since we already know that to this person, B is worse than another available alternative—namely, the alternative A). We do not expect to observe such a weird choice—but in some situations, this is exactly what has been observed; an example will be given later in this section.

This is a relatively clear example of a seemingly irrational decision making; we will show later that there are more subtle ones, where irrationality is not as easy to explain—but is clearly present; see examples in the next section.

So why irrational decision making? The fact that normal, reasonable people often make seemingly irrational decisions is puzzling. We humans come from billions years of improving evolution, we have flown to the Moon, we have discovered secrets of the Universe, we have learned to predict weather and to cure previously fatal diseases—this victorious image of Human Beings with a capital H does not seem to fit with simple decision mistakes, when the same person prefers A to B and then B to A, without even realizing that he/she is inconsistent.

Yes, we are not perfect, there are wars, crimes, and exploitations, but the common wisdom seems to indicate that most of our problems are caused by our selfishness—a criminal robs people because he wants to maximize his gain and he does not care if other people get hurt; a manufacturer of unhealthy foods maximizes his profit and does not care that people's health deteriorates as a result, etc. In all these cases, we blame the "evil" person for his selfish and vicious preferences, implicitly assuming that this person looks for what is best for him/her.

We can understand a person following a bad-for-society optimization criterion, but it is difficult to perceive a person whose decision making does not follow any optimization at all—and often, such a person is us.

How are such examples of seemingly irrational decision making explained now: the idea of bounded rationality. An established explanation for the seemingly irrational decision making is that we humans have a limited ability to process information—especially when the decision needs to be made urgently. On the qualitative level, this idea of *bounded rationality* is in good accordance with the observations; for example, usually, the more time we are given to make decisions, the more rational our decisions become; see, e.g., [61].

The existing explanations explain the very fact of seemingly irrational decision making, but not its observed specifics. The main limitation of the existing explanation is that it explains the *fact* that our decisions are sometimes not rational. In principle, under bounded resources, we can make different decisions, so we should observe various types of seemingly irrational decision making.

In many situations, however—for example, in the two situations described above—different decision makers exhibit the same deviations from the rational decision making. How can we explain these consistent deviations?

1.2 Examples of (Seemingly) Irrational Decision Making

Kahneman's book *Thinking, Fast and Slow* [61] has many examples of seemingly irrational decision making. In this book, we will concentrate on three examples. We selected these examples because they are, in our opinion, the easiest to explain without getting into the details of decision making theory and mathematical optimization. Let us describe these three examples in detail.

1.3   First Example of Seemingly Irrational Decision Making: Compromise Effect

This example is about shopping. The first example comes from simple decisions that most of us do very frequently: decisions on what to buy.

A customer shopping for an item usually has several choices. Some of these choices have better quality, leading to more possibilities, etc.—but are, on the other hand, more expensive. For example, a customer shopping for a photo camera has plenty of choices ranging from the cheapest ones (that work only in good lighting) to professional cameras that enable the users to make highest-quality photos even under complex circumstances. A traveller planning to spend a night in a new city has a choice from the cheapest motels (which provide a place to sleep) to luxurious hotels providing all kinds of comfort, etc. A customer selects one of the alternatives by taking into account the additional advantages of more expensive choices versus the need to pay more money for these choices.

In many real-life situations, customers face numerous choices. As usual in science, a good way to understand complex phenomena is to start by analyzing the simplest cases. In line with this reasoning, researchers provided customers with two alternatives and recorded which of these two alternatives a customer selected. In many particular cases, these experiments helped better understand the customer's selections—and sometimes even predict customer selections.

Expected behavior. At first glance, it seems like such pair-wise comparisons are all we need to know: if a customer faces several choices a_1, a_2, ..., a_n, then a customer will select an alternative a_i if and only if this alternative is better in pair-wise comparisons that all other possible choices.

An experiment. To confirm this common-sense idea, in the 1990s, several researchers asked the customers to select one of the three randomly selected alternatives.

The experimenters expected that since the three alternatives were selected at random, a customers would:

- sometimes select the cheapest of the three alternative (of lowest quality of all three),
- sometimes select the intermediate alternative (or intermediate quality), and
- sometimes select the most expensive of the three alternatives (of highest quality of all three).

Surprising results. Contrary to the expectations, the experimenters observed that in the overwhelming majority of cases, customers selected the intermediate alternative; see, e.g., [139, 143, 155]. In all these cases, the customer selected an alternative which provided a compromise between the quality and cost; because of this, this phenomenon was named *compromise effect*.

Why is this irrational? At first glance, selecting the middle alternative is reasonable. Let us show, however, that such a selection is not always rational.

For example, let us assume that we have four alternative $a_1 < a_2 < a_3 < a_4$ ordered in the increasing order of price and at the same time, increasing order of quality. Then:

- if we present the user with three choices $a_1 < a_2 < a_3$, in most cases, the user will select the middle choice a_2; this means, in particular, that, to the user, a_2 better than the alternative a_3;
- on the other hand, if we present the user with three other choices $a_2 < a_3 < a_4$, in most cases, the same user will select the middle choice a_3; but this means that, to the user, the alternative a_3 better than the alternative a_2.

If in a pair-wise comparison, a_2 is better, then the second choice is wrong. If in a pair-wise comparison, the alternative a_3 is better, then the first choice is wrong. In both cases, one of the two choices is irrational.

This is not just an experimental curiosity, customers' decisions have been manipulated this way. At first glance, the above phenomena may seem like one of optical illusions or logical paradoxes: interesting but not that critically important. Actually, it is serious and important, since, according to anecdotal evidence, many companies have tried to use this phenomenon to manipulate the customer's choices: to make the customer buy a more expensive product.

For example, if there are two possible types of a certain product, a company can make sure that most customers select the most expensive type—simply by offering, as the third option, an even more expensive type of the same product.

Manipulation possibility has been exaggerated, but mystery remains. Recent research showed that manipulation is not very easy: the compromise effect only happens when a customer has no additional information—and no time (or no desire) to collect such information. In situations when customers were given access to additional information, they selected—as expected from rational folks—one of the three alternatives with almost equal frequency, and their pairwise selections, in most cases, did not depend on the presence of any other alternatives; see, e.g., [148].

The new experiment shows that the compromise effect is not as critical and not as wide-spread as it was previously believed. However, in situation when decisions need to be made under major uncertainty, this effect is clearly present—and its seemingly counterintuitive, inconsistent nature is puzzling.

1.4   Second Example of Seemingly Irrational Decision Making: Biased Probability Estimates

More complicated decision making situations. In the first example, we considered situations with a simple choice, in which we have several alternatives, and we know the exact consequences of each alternative.

In many practical cases, the situation is more complicated: for each decision, depending on how things go, we may face different consequences. For example, if a person invests all his retirement money in the stock market, the market may go up—in which case, he will gain—or it may go down, in which case he will lose a big portion of his savings. A person who takes on a potentially dangerous sport (like car racing) will probably gain a lot of pleasure, but there is also a chance of a serious injury.

Need to estimate probabilities. To make a decision in such complex situations, it is important to estimate the probability of different outcomes. In some cases we know these probabilities—e.g., in the state-run lotteries, probabilities of winning are usually disclosed. In other cases, a person has to estimate these probabilities.

It seems reasonable to expect that people use unbiased estimates for probabilities. Of course, based on the limited information, we can get only approximate estimates of the corresponding probabilities. However, we expect that these estimates are unbiased, i.e., that, on average, they should provide a reasonably accurate estimate.

Indeed, if we systematically overestimate small probabilities, then we would overestimate our gain a lottery and, on average, lose. Similarly, if we systematically underestimate small probabilities, then, in particular, we will underestimate the probability of a disaster and invest in too risky stocks—and also lose on average.

Surprising observations. This is what we expect: unbiased estimates, but this is not what we observe. What we observe is that small probabilities are routinely *over*estimated, while probabilities close to 1 are routinely *under*estimated. This is not just an occasional phenomenon: for each actual probability, the estimated probability is consistently different. For different actual probabilities p_i, the corresponding estimated probabilities $\tilde{p}_i$ are given in [61] (see also references therein):

p_i	0	1	2	5	10	20	50	80	90	95	98	99	100
$\tilde{p}_i$	0	5.5	8.1	13.2	18.6	26.1	42.1	60.1	71.2	79.3	87.1	91.2	100

Why biased estimates? As we have mentioned, biased estimates are contrary to rational decision making: overestimating a small probability of success may get a decision maker involved in risky situations where, on average (and thus, in the long run) the decision maker will lose. On the other hand, overestimating a small probability of a disaster will make a decision maker too cautious and prevent him/her from making a rational risky decision.

1.5 Third Example of Seemingly Irrational Behavior: Average Instead of the Sum

Example. According to Chap. 15, Sect. 15.1 of [61], titled "Less Is More, Sometimes Even in Joint Evaluation", when pricing two large dinnerware sets,

- one consisting of 24 pieces in perfect condition, and
- the other consisting of the same 24 pieces plus 16 additional pieces, 9 of which are broken,

most people value the second set lower.

Why this is irrational. Rationally, this makes no sense, since after buying the second set, we can simply throw away the broken pieces, and actually end up with more pieces.

This example is a particular case of a general phenomenon. According to [61], this is a specific example of the following general phenomenon: when in a hurry, people often use an arithmetic average as a substitute for the sum; see, e.g., Chap. 8, Sect. 8.2 "Sets and Prototypes".

For example, when people are asked to compare two pictures with multiple line segments in each of them, and decide in which of the two pictures, the *total* length of its line segments is larger, they usually select the picture with the largest *average* length.

Need for an explanation. The general average-instead-sum phenomenon explains the dinnerware sets comparison—that people compare *average* values of pieces in two sets instead of comparing the overall values of these two sets: for the second set, the average value is indeed lower.

However, what needs explaining is the phenomenon itself, i.e., *why* average is used instead of the sum.

1.6 Structure of the Book

We start, in Chap. 2, with explaining how bounded rationality explains the above examples of seemingly irrational human behavior. Chapter 3 shows that bounded rationality can also explain other aspects of human decision making. In Chap. 4, we show that similar arguments *explain* the success of *heuristic techniques* in expert decision making. In Chap. 5, we use these explanations to *predict* the *quality* of the resulting *decisions*. Finally, in Chap. 6, we explain how we can *improve* on the existing *heuristic techniques* by formulating and solving the corresponding optimization problems. The final short Chap. 7 contains conclusions. Auxiliary computational issues are analyzed in the Appendix.

Chapter 2
Towards Explaining Specific Examples of Suboptimal (and Often Seemingly Irrational) Human Decisions

In this chapter, we show that bounded rationality indeed explains the seemingly irrational human decisions. Specifically, in Sect. 2.1, we briefly recall the traditional utility-based approach to decision making. In Sect. 2.2, we describe our main idea—of optimization under granularity. In Sect. 2.3, we show how this idea can explain the first example of seemingly irrational decision making: the compromise effect. In Sect. 2.4, we provide an explanation for the second example of seemingly irrational decision making: use of bias probability estimates. Finally, in Sect. 2.5, we explain why people use average instead of the sum.

2.1 Traditional Utility-Based Approach to Decision Making: A Brief Reminder

Main assumption behind the traditional decision theory. Traditional approach to decision making is based on an assumption that for each two alternatives A' and A'', a user can tell:

- whether the first alternative is better for him/her; we will denote this by $A'' < A'$;
- or the second alternative is better; we will denote this by $A' < A''$;
- or the two given alternatives are of equal value to the user; we will denote this by $A' = A''$.

Towards a numerical description of preferences: the notion of utility. Under the above assumption, we can form a natural numerical scale for describing preferences. Namely, let us select a very bad alternative A_0 and a very good alternative A_1. Then, most other alternatives are better than A_0 but worse than A_1.

For every probability $p \in [0, 1]$, we can form a lottery $L(p)$ in which we get A_1 with probability p and A_0 with probability $1 - p$.

© Springer International Publishing AG 2018
J. Lorkowski and V. Kreinovich, *Bounded Rationality in Decision Making Under Uncertainty: Towards Optimal Granularity*, Studies in Systems, Decision and Control 99, DOI 10.1007/978-3-319-62214-9_2

- When $p = 0$, this lottery coincides with the alternative A_0: $L(0) = A_0$.
- When $p = 1$, this lottery coincides with the alternative A_1: $L(1) = A_1$.

For values p between 0 and 1, the lottery is better than A_0 and worse than A_1. The larger the probability p of the positive outcome increases, the better the result:

$$p' < p'' \text{ implies } L(p') < L(p'').$$

Thus, we have a continuous scale of alternatives $L(p)$ that monotonically goes from $L(0) = A_0$ to $L(1) = A_1$. We will use this scale to gauge the attractiveness of each alternative A.

Due to the above monotonicity, when p increases, we first have $L(p) < A$, then we have $L(p) > A$, and there is a threshold separating values p for which $L(p) < A$ from the values p for which $L(p) > A$. This threshold value is called the *utility* of the alternative A:

$$u(A) \overset{\text{def}}{=} \sup\{p : L(p) < A\} = \inf\{p : L(p) > A\}.$$

Then, for every $\varepsilon > 0$, we have

$$L(u(A) - \varepsilon) < A < L(u(A) + \varepsilon).$$

We will describe such (almost) equivalence by $\equiv$, i.e., we will write that $A \equiv L(u(A))$.

How to elicit the utility from a user: a fast iterative process. Initially, we know the values $\underline{u} = 0$ and $\overline{u} = 1$ such that $A \equiv L(u(A))$ for some $u(A) \in [\underline{u}, \overline{u}]$.

On each stage of this iterative process, once we know values $\underline{u}$ and $\overline{u}$ for which $u(A) \in [\underline{u}, \overline{u}]$, we compute the midpoint u_{mid} of the interval $[\underline{u}, \overline{u}]$ and ask the user to compare A with the lottery $L(u_{\text{mid}})$ corresponding to this midpoint. There are two possible outcomes of this comparison: $A \leq L(u_{\text{mid}})$ and $L(u_{\text{mid}}) \leq A$.

- In the first case, the comparison $A \leq L(u_{\text{mid}})$ means that $u(A) \leq u_{\text{mid}}$, so we can conclude that $u \in [\underline{u}, u_{\text{mid}}]$.
- In the second case, the comparison $L(u_{\text{mid}}) \leq A$ means that $u_{\text{mid}} \leq u(A)$, so we can conclude that $u \in [u_{\text{mid}}, \overline{u}]$.

In both cases, after an iteration, we decrease the width of the interval $[\underline{u}, \overline{u}]$ by half. So, after k iterations, we get an interval of width 2^{-k} which contains $u(A)$—i.e., we get $u(A)$ with accuracy 2^{-k}.

How to make a decision based on utility values. Suppose that we have found the utilities $u(A')$, $u(A'')$,..., of the alternatives A', A'',...Which of these alternatives should we choose?

By definition of utility, we have:

- $A \equiv L(u(A))$ for every alternative A, and
- $L(p') < L(p'')$ if and only if $p' < p''$.

We can thus conclude that A' is preferable to A'' if and only if $u(A') > u(A'')$. In other words, we should always select an alternative with the largest possible value of utility. So, to find the best solution, we must solve the corresponding optimization problem.

Before we go further: caution. We are *not* claiming that people estimate probabilities when they make decisions: we know they often don't. Our claim is that when people make *definite* and *consistent* choices, these choices *can* be described by probabilities. (Similarly, a falling rock does not solve equations but follows Newton's equations $ma = m\dfrac{d^2x}{dt^2} = -mg$.) In practice, decisions are often *not* definite (uncertain) and *not* consistent.

How to estimate utility of an action. For each action, we usually know possible outcomes $S_1, \ldots, S_n$. We can often estimate the probabilities $p_1, \ldots, p_n$ of these outcomes.

By definition of utility, each situation S_i is equivalent to a lottery $L(u(S_i))$ in which we get:

- A_1 with probability $u(S_i)$ and
- A_0 with the remaining probability $1 - u(S_i)$.

Thus, the original action is equivalent to a complex lottery in which:

- first, we select one of the situations S_i with probability p_i: $P(S_i) = p_i$;
- then, depending on S_i, we get A_1 with probability $P(A_1 \mid S_i) = u(S_i)$ and A_0 with probability $1 - u(S_i)$.

The probability of getting A_1 in this complex lottery is:

$$P(A_1) = \sum_{i=1}^{n} P(A_1 \mid S_i) \cdot P(S_i) = \sum_{i=1}^{n} u(S_i) \cdot p_i.$$

In this complex lottery, we get:

- A_1 with probability $u = \sum_{i=1}^{n} p_i \cdot u(S_i)$, and
- A_0 with probability $1 - u$.

So, the utility of the complex action is equal to the sum u.

From the mathematical viewpoint, the sum defining u coincides with the expected value of the utility of an outcome. Thus, selecting the action with the largest utility means that we should select the action with the largest value of expected utility $u = \sum p_i \cdot u(S_i)$.

How uniquely determined is utility. The above definition of utility u depends on the selection of two fixed alternatives A_0 and A_1. What if we use different alternatives A_0' and A_1'? How will the new utility u' be related to the original utility u?

By definition of utility, every alternative A is equivalent to a lottery $L(u(A))$ in which we get A_1 with probability $u(A)$ and A_0 with probability $1 - u(A)$. For simplicity, let us assume that $A_0' < A_0 < A_1 < A_1'$. Then, for the utility u', we get $A_0 \equiv L'(u'(A_0))$ and $A_1 \equiv L'(u'(A_1))$. So, the alternative A is equivalent to a complex lottery in which:

- we select A_1 with probability $u(A)$ and A_0 with probability $1 - u(A)$;
- depending on which of the two alternatives A_i we get, we get A_1' with probability $u'(A_i)$ and A_0' with probability $1 - u'(A_i)$.

In this complex lottery, we get A_1' with probability

$$u'(A) = u(A) \cdot (u'(A_1) - u'(A_0)) + u'(A_0).$$

Thus, the utility $u'(A)$ is related with the utility $u(A)$ by a linear transformation $u' = a \cdot u + b$, with $a > 0$. In other words, utility is defined modulo a linear transformation.

Traditional approach summarized. We assume that

- we know possible actions, and
- we know the exact consequences of each action.

Then, we should select an action with the largest value of expected utility.

2.2 Our Main Idea: Optimization Under Granularity

When we do not have enough time to take all the information into account, a natural idea is to use partial information. For example, when a man sees an animal in the jungle, it could be a predator, so an immediate decision needs to be made on whether to run away or not. Ideally, we should take into account all the details of an animal image, but there is no time for that, a reasonable reaction is to run way if an animal is sufficiently large.

So, instead of considering each data set separately, we, in effect, combine these data sets into "granules" corresponding to the partial information that is actually used in decision making; see, e.g., [126]. In the above example, instead of using the animal's size, we only take into account whether this size is greater than a certain threshold s_0 or not. In effect, this means that we divide the set of all possible values of size into two granules:

- a granule consisting of small animals, whose size is smaller than s_0, and
- a granule consisting of large (and thus, potentially dangerous) animals, whose size is larger than or equal to s_0.

In this chapter, we show is that in many cases, if we take into account only algorithms that process such granular information, then the observed human decision making can be shown to be *optimal* among such granular algorithms—although, of course, if we could take into account all available information, we would be able to make a better decision.

2.3 Explaining the First Example of Seemingly Irrational Human Decision Making: Granularity Explains The Compromise Effect

In this section, we show that granularity explains the first example of seemingly irrational human decision making: the compromise effect. The results from this section first appeared in [85, 86, 90].

Compromise effect: reminder. We have three alternative a, a' and a'':

- the alternative a is the cheapest—and is, correspondingly, of the lowest quality among the given three alternatives;
- the alternative a' is intermediate in terms of price—and is, correspondingly, intermediate in terms of quality;
- finally, the alternative a'' is the most expensive—and is, correspondingly, of the highest quality among the given three alternatives.

We do not know the exact prices, we just know the order between them; similarly, we do not know the exact values of quality, we just know the order between them. In this situation, most people select an alternative a'.

Let us describe the corresponding granularity. The "utility" of each alternative comes from two factors:

- the first factor comes from the quality: the higher the quality, the better—i.e., larger the corresponding component u_1 of the utility;
- the second factor comes from price: the lower the price, the better for the user—i.e., the larger the corresponding component u_2 of the utility.

The fact that we do not know the exact value of the price means, in effect, that we consider three possible levels of price and thus, three possible levels of the utility u_1:

- low price, corresponding to high price-related utility;
- medium price, corresponding to medium price-related utility; and
- high price, corresponding to low price-related utility.

In the following text, we will denote "low" by L, "medium" by M, and "high" by H. In these terms, the above description of each alternative by the corresponding pair of utility values takes the following form:

- the alternative a is characterized by the pair (L, H);
- the alternative a' is characterized by the pair (M, M); and
- the alternative a'' is characterized by the pair (H, L).

Natural symmetries. We do not know a priori which of the two utility components is more important. As a result, it is reasonable to treat both components equally. In order words, the selection should be the same if we simply swap the two utility components—i.e., we should select the same of three alternatives before and after swap:

- if we are selecting an alternative based on the pairs (L, H), (M, M), and (H, L),
- then we should select the exact same alternative if the pairs were swapped, i.e., if:

  – the alternative a was characterized by the pair (H, L);
  – the alternative a' was characterized by the pair (M, M); and
  – the alternative a'' was characterized by the pair (L, H).

   Similarly, there is no reason to a priori prefer one alternative or the other. So, the selection should not depend on which of the alternatives we mark as a, which we mark as a', and which we mark as a''. In other words, any permutation of the three alternatives is a reasonable symmetry transformation. For example, if, in our case, we select an alternative a which is characterized by the pair (L, H), then, after we swap a and a'' and get the choice of the following three alternatives:

- the alternative a which is characterized by the pair (H, L);
- the alternative a' is characterized by the pair (M, M); and
- the alternative a'' is characterized by the pair (L, H),

then we should select the same alternative—which is now denoted by a''.

General comment: symmetries have been helpful in dealing with uncertainty. It should be mentioned that in situations with major uncertainty, symmetries are often helpful. The main idea behind using symmetries is that if the situation is invariant with respect to some natural symmetries, then it is reasonable to select an action which is also invariant with respect to all these symmetries.

   There have been many applications of this idea, starting from the pioneering work of N. Wiener on Cybernetics; see, e.g., [161]. It has been shown that for many empirically successful techniques related to neural fuzzy logic, networks, and interval computations, their empirical success can be explained by the fact that these techniques can be deduced from the appropriate symmetries; see, e.g., [115]. In particular, this explains the use of a sigmoid activation function $s(z) = \dfrac{1}{1 + \exp(-z)}$ in neural networks, the use of the most efficient "and"-operations (t-norms) and "or"-operations (t-conorms) in fuzzy logic, etc. [115].

Back to our situation: what we can conclude based on the corresponding symmetries. One can observe the following: that if we *both* swap u_1 and u_2 *and* swap a and a'', then you get the exact same characterization of all alternatives:

- the alternative a is still characterized by the pair (L, H);
- the alternative a' is still characterized by the pair (M, M); and
- the alternative a'' is still characterized by the pair (H, L).

The only difference is that:

- now, a indicates an alternative which was previously denoted by a'', and
- a'' now denotes the alternative which was previously denoted by a.

   As we have mentioned, it is reasonable to conclude that:

- if in the original triple selection, we select the alternative a,

- then in the new selection—which is based on the exact same pairs of utility values—we should also select an alternative denoted by a.

But this "new" alternative a is nothing else but the old a''. So, we conclude that:

- if we selected a,
- then we should have selected a different alternative a'' in the original problem.

This is clearly a contradiction:

- we started by assuming that, to the user, a was better than a'' (because otherwise a would not have been selected in the first place), and
- we ended up concluding that to the same user, the original alternative a'' is better than a.

This contradiction shows that, under the symmetry approach, we cannot prefer a.
   Similarly:

- if in the original problem, we preferred an alternative a'',
- then this would mean that in the new problem, we should still select an alternative which marked by a''.

But this "new" a'' is nothing else but the old a. So, this means that:

- if we originally selected a'',
- then we should have selected a different alternative a in the original problem.

This is also a contradiction:

- we started by assuming that, to the user a'' was better than a (because otherwise a'' would not have been selected in the first place), and
- we ended up concluding that to the same user, the original alternative a is better than a''. This contradiction shows that, under the symmetry approach, we cannot prefer a''.

We thus conclude that out of the three alternatives a, a', and a'':

- we cannot select a, and
- we cannot select a''.

This leaves us only once choice: to select the intermediate alternative a'.
   This is exactly the compromise effect that we planned to explain.

Conclusion. Experiments show when people are presented with three choices $a < a' < a''$ of increasing price and increasing quality, and they do not have detailed information about these choices, then in the overwhelming majority of cases, they select the intermediate alternative a'.

   This "compromise effect" is, at first glance, irrational: selecting a' means that, to the user, a' is better than a'', but in a similar situation when the user is presented with $a' < a'' < a'''$, the same principle would indicate that the user will select a''—meaning that a'' is better than a'.

   Somewhat surprisingly, a natural granularity approach explains this seemingly irrational decision making.

2.4 Explaining the Second Example of Seemingly Irrational Human Decision Making: Granularity Explains Why Our Probability Estimates Are Biased

In this section, we show that granularity explains the second example of seemingly irrational human decision making: that in our decisions, we use biased estimates of probabilities.

The results from this section appeared in [82, 86, 90].

Main idea. Probability of an event is estimated, from observations, as the frequency with which this event occurs. For example, if out of 100 days of observation, rain occurred in 40 of these days, then we estimate the probability of rain as 40%. In general, if out of n observations, the event was observed in k of them, we estimate the probability as the ratio $\dfrac{k}{n}$.

This ratio is, in general, different from the actual (unknown) probability. For example, if we take a fair coin, for which the probability of head is exactly 50%, and flip it 100 times, we may get 50 heads, but we may also get 47 heads, 52 heads, etc. Similarly, if we have the coin fall heads 50 times out of 100, the actual probability could be 50%, could be 47% and could be 52%. In other words, instead of the exact value of the probability, we get a *granule* of possible values. (In statistics, this granule is known as a *confidence interval*; see, e.g., [144].)

In other words:

- first, we estimate a probability based on the observations; as a result, instead of the exact value, we get a granule which contains the actual (unknown) probability; this granule is all we know about the actual probability;
- then, when a need comes to estimate the probability, we produce an estimate based on the granule.

Let us analyze these two procedures one by one.

Probability granules: analysis of the first procedure and the resulting formulas. It is known (see, e.g., [144]), that the expected value of the frequency is equal to p, and that the standard deviation of this frequency is equal to

$$\sigma = \sqrt{\frac{p \cdot (1 - p)}{n}}.$$

It is also known that, due to the Central Limit Theorem, for large n, the distribution of frequency is very close to the normal distribution (with the corresponding mean p and standard deviation σ).

For normal distribution, we know that with a high certainty all the values are located within 2–3 standard deviations from the mean, i.e., in our case, within the interval $(p - k_0 \cdot \sigma, \ p + k_0 \cdot \sigma)$, where $k_0 = 2$ or $k_0 = 3$: for example, for $k_0 = 3$, this is true with confidence 99.9%. We can thus say that the two values of probability p and p' are (definitely) distinguishable if the corresponding intervals of possible values of frequency do not intersect—and thus, we can distinguish between these two probabilities just by observing the corresponding frequencies.

In precise terms, the probabilities $p < p'$ are distinguishable if

$$(p - k_0 \cdot \sigma, p + k_0 \cdot \sigma) \cap (p' - k_0 \cdot \sigma', p' + k_0 \cdot \sigma') = \emptyset,$$

where

$$\sigma' \stackrel{\text{def}}{=} \sqrt{\frac{p' \cdot (1 - p')}{n}},$$

i.e., if $p' - k_0 \cdot \sigma' \geq p + k_0 \cdot \sigma$. The smaller p', the smaller the difference $p' - k_0 \cdot \sigma'$. Thus, for a given probability p, the next distinguishable value p' is the one for which

$$p' - k_0 \cdot \sigma' = p + k_0 \cdot \sigma.$$

When n is large, these value p and p' are close to each other; therefore, $\sigma' \approx \sigma$. Substituting an approximate value σ instead of σ' into the above equality, we conclude that

$$p' \approx p + 2k_0 \cdot \sigma = p + 2k_0 \cdot \sqrt{\frac{p \cdot (1 - p)}{n}}.$$

If the value p corresponds to the i-th level, then the next value p' corresponds to the $(i + 1)$-st level. Let us denote the value corresponding to the i-th level by $p(i)$. In these terms, the above formula takes the form

$$p(i + 1) - p(i) = 2k_0 \cdot \sqrt{\frac{p \cdot (1 - p)}{n}}.$$

The above notation defines the value $p(i)$ for non-negative integers i. We can extrapolate this dependence so that it will be defined for all non-negative real values i.

When n is large, the values $p(i + 1)$ and $p(i)$ are close, the difference

$$p(i + 1) - p(i)$$

is small, and therefore, we can expand the expression $p(i + 1)$ in Taylor series and keep only linear terms in this expansion:

$$p(i + 1) - p(i) \approx \frac{dp}{di}.$$

Substituting the above expression for $p(i + 1) - p(i)$ into this formula, we conclude that

$$\frac{dp}{di} = \text{const} \cdot \sqrt{p \cdot (1 - p)}.$$

Moving all the terms containing p into the left-hand side and all the terms containing i into the right-hand side, we get

$$\frac{dp}{\sqrt{p \cdot (1-p)}} = \text{const} \cdot di.$$

Integrating this expression and taking into account that $p = 0$ corresponds to the lowest 0-th level—i.e., that $i(0) = 0$—we conclude that

$$i(p) = \text{const} \cdot \int_0^p \frac{dq}{\sqrt{q \cdot (1-q)}}.$$

This integral can be easily computed if introduce a new variable t for which $q = \sin^2(t)$. In this case,

$$dq = 2 \cdot \sin(t) \cdot \cos(t) \cdot dt,$$

$1 - p = 1 - \sin^2(t) = \cos^2(t)$ and therefore,

$$\sqrt{p \cdot (1-p)} = \sqrt{\sin^2(t) \cdot \cos^2(t)} = \sin(t) \cdot \cos(t).$$

The lower bound $q = 0$ corresponds to $t = 0$ and the upper bound $q = p$ corresponds to the value t_0 for which $\sin^2(t_0) = p$—i.e., $\sin(t_0) = \sqrt{p}$ and $t_0 = \arcsin\left(\sqrt{p}\right)$. Therefore,

$$i(p) = \text{const} \cdot \int_0^p \frac{dq}{\sqrt{q \cdot (1-q)}} = \text{const} \cdot \int_0^{t_0} \frac{2 \cdot \sin(t) \cdot \cos(t) \cdot dt}{\sin(t) \cdot \cos(t)} =$$

$$\int_0^{t_0} 2 \cdot dt = 2 \cdot \text{const} \cdot t_0.$$

We know how t_0 depends on p, so we get

$$i(p) = 2 \cdot \text{const} \cdot \arcsin\left(\sqrt{p}\right).$$

We can determine the constant from the condition that the largest possible probability value $p = 1$ should correspond to the largest level $i = m$. From the condition that $i(1) = m$, taking into account that

$$\arcsin\left(\sqrt{1}\right) = \arcsin(1) = \frac{\pi}{2},$$

we conclude that

$$i(p) = \frac{2m}{\pi} \cdot \arcsin\left(\sqrt{p}\right).$$

Thus,

$$\arcsin\left(\sqrt{p}\right) = \frac{\pi \cdot i}{2m},$$

hence

$$\sqrt{p} = \sin\left(\frac{\pi \cdot i}{2m}\right)$$

and thus,

$$p(i) = \sin^2\left(\frac{\pi \cdot i}{2m}\right).$$

Thus, probability granules are formed by intervals $[p(i), p(i+1)]$. Each empirical probability is represented by the granule i to which it belongs.

From granules to probability estimates: analysis of the second procedure. As we have mentioned, instead of the actual probabilities, we have probability *labels* corresponding to m different granules:

- the first label corresponds to the smallest certainty,
- the second label corresponds to the second smallest certainty,
- etc.,
- until we reach the last label which corresponds to the largest certainty.

We need to produce some estimates of the probability based on the granule. In other words, for each i from 1 to m, we need to assign, to each i-th label, a value p_i in such a way that labels corresponding to higher certainty should get larger numbers: $p_1 < p_2 < \cdots < p_m$.

Before we analyze how to do it, let us recall that one of the main objectives of assigning numerical values is that we want computers to help us solve the corresponding decision problems, and computers are not very good in dealing with granules; their natural language is the language of numbers. From this viewpoint, it makes sense to consider not all theoretically possible exact real numbers, but only computer-representable real numbers.

In a computer, real numbers from the interval $[0, 1]$ are usually represented by the first d digits of their binary expansion. Thus, computer-representable numbers are $0, h \stackrel{\text{def}}{=} 2^{-d}, 2h, 3h, \ldots$, until we reach the value $2^d \cdot h = 1$.

In our analysis, we will assume that the "machine unit" $h > 0$ is fixed, and we will thus assume that only multiples of this machine units are possible values of all n probabilities p_i.

For example, when $h = 0.1$, each probability p_i takes 11 possible values: 0, 0.1, 0.2, 0.3, 0.4, 0.5, 0.6, 0.7, 0.8, 0.9, and 1.0.

In the modern computers, the value h is extremely small; thus, whenever necessary, we can assume that $h \approx 0$—i.e., use limit case of $h \to 0$ instead of the actual small "machine unit" h.

For each h, we consider all possible combinations of probabilities $p_1 < \cdots < p_m$ in which all the numbers p_i are proportional to the selected step h, i.e., all possible combinations of values $(k_1 \cdot h, \ldots, k_m \cdot h)$ with $k_1 < \ldots < k_m$.

For example, when $m = 2$ and $h = 0.1$, we consider all possible combinations of values $(k_1 \cdot h, k_2 \cdot h)$ with $k_1 < k_2$:

- For $k_1 = 0$ and $p_1 = 0$, we have 10 possible combinations $(0, 0.1)$, $(0, 0.2)$,..., $(0, 1)$.
- For $k_1 = 1$ and $p_1 = 0.1$, we have 9 possible combinations $(0.1, 0.2)$, $(0.1, 0.3)$,..., $(0.1, 1)$.
- ...
- Finally, for $k_1 = 9$ and $p_1 = 0.9$, we have only one possible combination $(0.9, 1)$.

For each i, for different possible combinations $(p_1, \ldots, p_m)$, we get, in general, different value of the probability p_i. According to the complete probability formula, we can obtain the actual (desired) probability P_i if we combine all these value p_i with the weights proportional to the probabilities of corresponding combinations:

$$P_i = \sum_{p_1 < \cdots < p_m} p_i \cdot \text{Prob}(p_1, \ldots, p_m).$$

Since we have no reason to believe that some combinations $(p_1, \ldots, p_m)$ are more probable and some are less probable, it is thus reasonable to assume that all these combinations are equally probable. Hence, P_i is equal to the arithmetic average of the values p_i corresponding to all possible combinations $(p_1, \ldots, p_m)$.

For example, for $m = 2$ and $h = 0.1$, we thus estimate P_1 by taking an arithmetic average of the values p_1 corresponding to all possible pairs. Specifically, we average:

- ten values $p_1 = 0$ corresponding to ten pairs $(0, 0.1)$,..., $(0, 1)$;
- nine values $p_1 = 0.1$ corresponding to nine pairs $(0.1, 0.2)$,..., $(0.1, 1)$;
- ...
- and a single value $p_1 = 0.9$ corresponding to the single pair $(0.9, 1)$.

As a result, we get the value

$$P_1 = \frac{10 \cdot 0.0 + 0 \cdot 0.1 + \cdots + 1 \cdot 0.9}{10 + 9 + \cdots + 1} = \frac{16.5}{55} = 0.3.$$

Similarly, to get the value p_2, we average:

- a single value $p_2 = 0.1$ corresponding to the single pair $(0, 0.1)$;
- two values $p_2 = 0.2$ corresponding to two pairs $(0, 0.2)$ and $(0.1, 0.2)$;
- ...
- ten values $p_2 = 1.0$ corresponding to ten pairs $(0, 1)$, ..., $(0.9, 1)$.

As a result, we get the value

$$P_2 = \frac{1 \cdot 0.1 + 2 \cdot 0.2 + \cdots + 10 \cdot 1.0}{1 + 2 + \cdots + 10} = \frac{37.5}{55} = 0.7.$$

The probability p_i of each label can take any of the equidistant values $0, h, 2h, 3h, \ldots$, with equal probability. In the limit $h \to 0$, the resulting probability distribution tends to the uniform distribution on the interval $[0, 1]$.

In this limit $h \to 0$, we get the following problem:

- we start with m independent random variables $v_1, \ldots, v_m$ which are uniformly distributed on the interval $[0, 1]$;
- we then need to find, for each i, the conditional expected value

$$E[v_i \mid v_1 < \cdots < v_m]$$

of each variable v_i under the condition that the values v_i are sorted in increasing order.

Conditional expected values are usually more difficult to compute than unconditional ones. So, to solve our problem, let us reduce our problem to the problem of computing the usual (unconditional) expectation.

Let us consider m independent random variables each of which is uniformly distributed on the interval $[0, 1]$. One can easily check that for any two such variables v_i and v_j, the probability that they are equal to each other is 0. Thus, without losing generality, we can safely assume that all m random values are different. Therefore, the whole range $[0, 1]^m$ is divided into $m!$ sub-ranges corresponding to different orders between v_i. Each sub-range can be reduced to the sub-range corresponding to $v_1 < \cdots < v_m$ by an appropriate permutation in which v_1 is swapped with the smallest $v_{(1)}$ of m values, v_2 is swapped with the second smallest $v_{(2)}$, etc.

Thus, the conditional expected value of v_i is equal to the (unconditional) expected value of the i-th value $v_{(i)}$ in the increasing order. This value $v_{(i)}$ is known as an *order statistic*, and for uniform distributions, the expected values of all order statistics are known (see, e.g., [1, 4, 25]): $P_i = \dfrac{i}{m + 1}$.

So, if all we know is that our degree of certainty is expressed by i-th label on an m-label scale of granules, then it is reasonable to assign, to this case, the probability $P_i = \dfrac{i}{m + 1}$.

Let us now combine the two procedures. In the first procedure, based on the empirical frequency p, we find a label i for which

$$p \approx \sin^2\left(\frac{\pi \cdot i}{2m}\right).$$

Based on this label, we then estimate the probability as $P_i = \dfrac{i}{m + 1}$. For large m, we have $P \approx \dfrac{i}{m}$. Substituting P instead of $\dfrac{i}{m}$ into the formula for p, we conclude that

$$p \approx \sin^2\left(\frac{\pi}{2} \cdot P\right).$$

Based on this formula, we can express the estimate P in terms of the actual probability p, as

$$P \approx \frac{1}{\pi} \cdot \arcsin(\sqrt{p}).$$

Comparing our estimates P with empirical probability estimates $\widetilde{p}_i$: first try. Let us compare the probabilities p_i, Kahneman's empirical estimates $\widetilde{p}_i$, and the estimates $P_i = \frac{1}{\pi} \cdot \arcsin(\sqrt{p_i})$ computed by using the above formula:

p_i	0	1	2	5	10	20	50	80	90	95	98	99	100
$\widetilde{p}_i$	0	5.5	8.1	13.2	18.6	26.1	42.1	60.1	71.2	79.3	87.1	91.2	100
P_i	0	6.4	9.0	14.4	20.5	29.5	50.0	70.5	79.5	85.6	91.0	93.6	100

For most probabilities p_i, the difference between the values P_i' and the empirical probability estimates $\widetilde{p}_i$ is so small that it is below the accuracy with which the empirical weights can be obtained from the experiment.

Thus, granularity ideas indeed explain Kahneman and Tversky's observation of biased empirical probability estimates.

Summary. Kahneman and Tversky showed that when people make decisions, then instead of—as should be rational—weighting outcomes with weights proportional to probabilities of different outcomes—they use *biased* weights, overestimating the importance of low-probability events and underestimating the importance of high-probability events. In this section, we show that this observable bias can be explained if we take into account granularity—imposed by our limited rationality (i.e., our limited ability to process information).

2.5 Explaining the Third Example of Seemingly Irrational Human Decision Making: Using Average Instead of the Sum

In this section, we show how to explain the third example of seemingly irrational behavior: using average instead of the sum. The result of this section first appeared in [76].

Using average instead of the sum: reminder. According to [61], when in a hurry, people often use an arithmetic average as a substitute for the sum. This substitution leads to a seemingly irrational behavior. For example, when pricing two large dinnerware sets,

• one consisting of 24 pieces in perfect condition, and
• the other consisting of the same 24 pieces plus 16 additional ones, 9 of which are broken,

most people value the second set lower—possibly because in the second set, the average value of a piece is lower. This selection is irrational, since after buying the second set, we can simply throw away the broken pieces, and actually end up with more pieces.

Why do people use the average instead of the sum: a possible explanation. Our explanation for the use of arithmetic averages is that the arithmetic average is much easier to compute than, e.g., the sum.

This may sound somewhat counter-intuitive, because, at first glance, the formula for the arithmetic average $\overline{x} = \dfrac{x_1 + \cdots + x_n}{n}$ looks somewhat more complex to compute than the formulas for the sum $s = x_1 + \cdots + x_n$: to compute the average, we need to perform all the additions needed for the sum plus one additional division.

This is indeed the case if we talk about *exact* computations: to compute the exact sum or the exact average, one needs to process each of n numbers at least once—if we do not process one of the numbers, we cannot get the exact value of sum or average. Since each elementary arithmetic operation takes at most two numerical inputs, this means that in both cases, we need at least $n/2$ operations, and $n/2$ is $O(n)$.

However, if we take into account that the values x_i are only known approximately and that, as a result, we only need approximate values of sum and average, then the computational complexity changes. For the sum, we still need to count the intervals, but to compute the approximate values of the average, we can use Monte-Carlo techniques: namely, we can select a random sample of values and take the arithmetic average of this sample.

According to the Large Numbers Theorem, when the sample size is large, this random-sample-based arithmetic average provides a good approximation to the desired exact average—and the larger the sample, the more accurate this approximation; see, e.g., [144].

The required sample size—and thus, the corresponding computational complexity of estimating the average this way—depends only on the desired accuracy of estimating the average, and does not depend on the number n of original values. Thus, for a fixed accuracy, the computational complexity of this algorithm does not grow with n at all, it is $O(1)$, while the complexity of computing the sum still grows with n as $O(n)$. Since for large n, $O(1) \ll O(n)$, this explains why people use an average as a substitute for the sum.

Chapter 3
Towards Explaining Other Aspects of Human Decision Making

In the previous chapter, we explained how bounded rationality ideas can explain seemingly irrational human decisions. In this chapter, we show that it can explain other not-easy-to-explain aspects of human decision making.

As we mentioned in Sect. 2.1, the decision making theory shows that a rational decision is the one that maximizes the expected utility. Thus, to make a decision, we need to estimate the utilities, we need to estimate the probabilities, and we need to make a decision based on these estimates. These aspects of human decision making are discussed, correspondingly, in Sects. 3.1, 3.2, and 3.3.

The corresponding issues are related to individual decision making. Of course, in real life, when we make decisions, we need to take into account the utility of others. A general aspect of this taking-into-account is discussed in Sect. 3.4. In the last Sect. 3.5, we pay special attention to decision making situations in which the utility of others is the main concern: namely, the situation of education: its first subsection deals with the "how" aspects of education, and its second subsection analyzes the results of education.

3.1 Estimating Utility: Why Utility Grows as Square Root of Money

Estimating utility is important. Since the optimal decision making is equivalent to maximizing expected utility, to make good recommendations, we need to know the utility of different outcomes.

Empirical fact. It has been experimentally determined that for situations with monetary gain, utility u grows with the money amount x as $u \approx x^{\alpha}$, with $\alpha \approx 0.5$, i.e., approximately as $u \approx \sqrt{x}$; see, e.g., [61] and references therein.

What we do in this section. In this section, we provide another example when granularity explains observed decision making. Namely, we explain why utility that

© Springer International Publishing AG 2018
J. Lorkowski and V. Kreinovich, *Bounded Rationality in Decision Making Under Uncertainty: Towards Optimal Granularity*, Studies in Systems, Decision and Control 99, DOI 10.1007/978-3-319-62214-9_3

describes human decision making grows approximately as square root of money amount.

The results from this section first appeared in [90].

Main idea behind our explanation. Money is useful because one can buy goods and services with it. The more goods and services one buys, the better. In the first approximation, we can say that the utility increases with the increase in the number of goods and service.

In these terms, to estimate the utility corresponding to a given amount of money, we need to do two things:

- first, we need to estimate how many goods and services a person can buy for a given amount of money;
- second, we need to estimate what value of utility corresponds to this number of goods and services.

Step 1: estimating how many goods and services a person can buy. Different goods and services have different costs c_i; some are cheaper, some are more expensive. We know that all the costs c_i are bounded by some reasonable number C, so they are all located within an interval $[0, C]$. Let us sort the costs of different items in increasing order: $c_1 < c_2 < \cdots < c_n$.

In these terms, the smallest amount of money that we need to buy a single item is c_1. The smallest amount of money that we need to buy two items is $c_1 + c_2$, etc. In general, the smallest amount of money that we need to buy k items is $c_1 + c_2 + \cdots + c_k$.

How does this amount depend on k? We do not know the exact costs c_i, all we know is that these costs are sorted in increasing order. Similarly to the previous section, we can therefore consider all possible combinations $c_1 < \cdots < c_n$, and take, as an estimate C_i for c_i, the average value of c_i over all such combinations. Similarly to the previous section, we can conclude that $C_i = C \cdot \dfrac{i}{n+1}$.

In these terms, the expected amount of money needed to buy k items is equal to

$$C_1 + C_2 + \cdots + C_k = \frac{C}{n} \cdot (1 + 2 + \cdots + k) = \frac{C}{2n} \cdot k \cdot (k+1) \approx \text{const} \cdot k^2.$$

Step 2: estimating the utility corresponding to k items. Let u_k denote the utility corresponding to k items. We know that all the values u_k are bounded by some reasonable number U, so they are all located within an interval $[0, U]$. Clearly the more items, the better, i.e., the larger utility. Thus, we conclude that $u_1 < u_2 < \cdots < u_n$.

We do not know the exact values of u_k, all we know is that these utility values are sorted in increasing order. We can thus consider all possible combinations $u_1 < \cdots < u_n$, and take, as an estimate U_k for u_k, the average value of u_k over all such combinations. Similarly to the previous section, we can conclude that $U_k = U \cdot \dfrac{k}{n+1} = \text{const} \cdot k$.

Let us combine these two estimates. What is the utility corresponding to the amount of money x? To answer this question, first, we estimate the number of items k that we can buy with this amount. According to our estimates, $x = \text{const} \cdot k^2$, so we conclude that $k = \text{const} \cdot \sqrt{x}$. Then, we use this value k to estimate the utility $U \approx U_k$. Substituting $k = \text{const} \cdot \sqrt{x}$ into the formula $U \approx U_k = \text{const} \cdot k$, we conclude that $U \approx \text{const} \cdot \sqrt{x}$.

Since, as we have mentioned, utility is defined modulo a linear transformation, we can thus conclude that $U \approx \sqrt{x}$, which is exactly what we wanted to explain.

Summary. Thus, granularity indeed explains an interesting difficult-to-explain empirical fact—that utility grows as square root of money amount.

3.2   Estimating Probabilities: A Justification of Sugeno λ-Measures

To describe expert uncertainty, researchers and practitioners often go beyond (additive) probability measures and use non-additive (fuzzy) measures. One of the most widely used classes of such measures is the class of Sugeno λ-measures. Their success is somewhat paradoxical, since from the purely mathematical viewpoint, these measures are—in some reasonable sense—equivalent to probability measures. In this section, we explain this success by showing that while *mathematically*, it is possible to reduce Sugeno measures to probability measures, but from the *computational* viewpoint, using Sugeno measures is much more efficient. We also show that among all fuzzy measures which are equivalent to probability measures, Sugeno measures (and a slightly more general family of measures) are the only ones with this efficiency property.

Results from this section first appeared in [117].

Traditional approach: probability measures. Usually, uncertainty has been described by probabilities. In mathematical terms, probabilistic information about events from some set X of possible events is usually described in terms of a *probability measure*, i.e., a function $p(A)$ that maps some sets $A \subseteq X$ into real numbers from the interval $[0, 1]$.

The probability $p(A)$ of a set A is usually interpreted as the frequency with which events from the set A occur in real life. In this interpretation, if we have two disjoint sets A and B with $A \cap B = \emptyset$, then the frequency $p(A \cup B)$ with which the events from A or B happen is equal to the sum of the frequencies $p(A)$ and $p(B)$ corresponding to each of these sets.

This property of probabilities measures is known as *additivity*: if $A \cap B = \emptyset$, then

$$p(A \cup B) = p(A) + p(B). \tag{3.2.1}$$

Need to do beyond probability measures. Since the appearance of fuzzy sets (see, e.g., [65, 120, 163]), it has been argued that to adequately describe expert knowledge, we often need to go beyond probabilities. In general, instead of probabilities, we have the expert's *degree of confidence* $g(A)$ that an event from the set A will actually occur.

Clearly, something should occur, so $g(\emptyset) = 0$ and $g(X) = 1$. Also, it is reasonable to take into account that the larger the set, the more confident we are that an event from this set will occur, i.e., $A \subseteq B$ implies $g(A) \leq g(B)$. Functions $g(A)$ that satisfy these properties are known as *fuzzy measures*.

Sugeno λ-measures. M. Sugeno, one of the pioneers of fuzzy measures, introduced a specific class of fuzzy measures which are now known as *Sugeno λ-measures* [151]. Measures from this class are close to the probability measures in the following sense: similarly to the case of probability measures, if we know $g(A)$ and $g(B)$ for two disjoint sets, we can still reconstruct the degree $g(A \cup B)$. The difference is that this reconstructed value is no longer the sum $g(A) + g(B)$, but a slightly more complex expression.

To be more precise, Sugeno λ-measures satisfy the following property: if $A \cap B = \emptyset$, then

$$g(A \cup B) = g(A) + g(B) + \lambda \cdot g(A) \cdot g(B), \qquad (3.2.2)$$

where $\lambda > -1$ is a real-valued parameter.

When $\lambda = 0$, the formula (3.2.2) corresponding to the Sugeno measure transforms into the additivity formula (3.2.1) corresponding to the probability measure. From this viewpoint, the value λ describes how close the given Sugeno measure is to a probability measure: the smaller $|\lambda|$, the closer these measures are.

Sugeno λ-measures had many practical applications. Sugeno measures are among the most widely used fuzzy measures; see, e.g., [12, 153, 158] and references therein.

Comment. Of course, not all expert reasoning can be described by Sugeno measures. In many practical situations, the expert's degree of confidence $g(A \cup B)$ in the event $A \cup B$ is *not* uniquely determined by the values $g(A)$ and $g(B)$. In such situations, we need to consider more general classes of fuzzy measures [12, 153, 158].

Problem. This practical success is somewhat paradoxical. Indeed:

- The main point of using fuzzy measures is to go beyond probability measures.
- On the other hand, Sugeno λ-measures are, in some reasonable sense, equivalent to probability measures (see [111] and the following text).

How can we explain this?

What we do in this section. In this section, we explain the seeming paradox of Sugeno λ-measures as follows:

- Yes, from the purely mathematical viewpoint, Sugeno measures are indeed equivalent to probability measures.
- However, from the computational viewpoint, processing Sugeno measure directly is much more computationally efficient than using a reduction to a probability measure.

We also analyze which other probability-equivalent fuzzy measures have this property: it turns out that this property holds only for Sugeno measures themselves and for a slightly more general class of fuzzy measures.

The structure of this section is straightforward: first, following the main ideas from [111], we describe in what sense Sugeno measure is mathematically equivalent to a probability measure. Then, we explain why processing Sugeno measures is more computationally efficient than using a reduction to probabilities, and finally, we analyze what other fuzzy measures have this property.

What we mean by equivalence. According to the formula (3.2.2), if we know the values $a = g(A)$ and $b = g(B)$ for disjoint sets A and B, then we can compute the value $c = g(A \cup B)$ as

$$c = a + b + \lambda \cdot a \cdot b. \tag{3.2.3}$$

We would like to find a 1-1 function $f(x)$ for which $p(A) \overset{\text{def}}{=} f^{-1}(g(A))$ is a probability measure, i.e., for which, if c is obtained by the relation (3.2.3), then for the values

$$a' = f^{-1}(a), \quad b' = f^{-1}(b), \quad \text{and} \quad c' = f^{-1}(c),$$

we should have

$$c' = a' + b'.$$

Comment. As we have mentioned, $A \subseteq B$ implies both $p(A) \leq p(B)$ and $g(A) \leq g(B)$. Thus, larger probability values should lead to larger degrees of confidence. It is therefore reasonable to also require that the mapping $f(x)$ that transforms the probability $p(A)$ into the corresponding degree of confidence $g(A) = f(p(A))$ be monotonic. It should be mentioned that for continuous functions $f(x)$, monotonicity automatically follows from our requirement that f is a 1-1 function.

How to show that a Sugeno λ-measure with $\lambda \neq 0$ is equivalent to a probability measure. Let us consider the auxiliary values $A = 1 + \lambda \cdot a$, $B = 1 + \lambda \cdot b$, and $C = 1 + \lambda \cdot c$. From the formula (3.2.3), we can now conclude that

$$C = 1 + \lambda \cdot (a + b + \lambda \cdot a \cdot b) = 1 + \lambda \cdot a + \lambda \cdot b + \lambda^2 \cdot a \cdot b. \tag{3.2.4}$$

One can easily check that the right-hand side of this formula is equal to the product $A \cdot B$ of the expressions $A = 1 + \lambda \cdot a$ and $B = 1 + \lambda \cdot b$. Thus, we get

$$C = A \cdot B. \tag{3.2.5}$$

We have a product, we need a sum. Converting from the product to the sum is easy: it is known that logarithm of the product is equal to the sum of logarithms. Thus, for the values

$$a' = \ln(A) = \ln(1 + \lambda \cdot a),$$

$$b' = \ln(B) = \ln(1 + \lambda \cdot b),$$

and
$$c' = \ln(C) = \ln(1 + \lambda \cdot c),$$

we get the desired formula
$$c' = a' + b'.$$

To get this formula, we used the inverse transformation f^{-1} that transforms each value x into a new value
$$x' = \ln(1 + \lambda \cdot x). \tag{3.2.6}$$

When $\lambda > 0$, then for $x \geq 0$, we get $1 + \lambda \cdot x \geq 1$ and thus, $x' = \ln(1 + \lambda \cdot x) \geq 0$.

When $\lambda < 0$, then for $x > 0$, we have $1 + \lambda \cdot x < 1$ and thus, $x' = \ln(1 + \lambda \cdot x) < 0$. However, we want to interpret the values x' as probabilities, and probabilities are always non-negative. Therefore, for $\lambda < 0$, we need to change the sign and consider

$$x' = -\ln(1 + \lambda \cdot x). \tag{3.2.7}$$

For these new values, (3.2.3) still implies that $c' = a' + b'$.

From the relations (3.2.6) and (3.2.7), we can easily find the corresponding direct transformation $x = f(x')$. Indeed, for $\lambda > 0$, by exponentiating both sides of the formula (3.2.6), we get $1 + \lambda \cdot x = \exp(x')$, hence

$$f(x') = \frac{1}{\lambda} \cdot (\exp(x') - 1). \tag{3.2.8}$$

For $\lambda < 0$, by exponentiating both sides of the formula (3.2.7), we get $1 + \lambda \cdot x = \exp(-x')$, hence
$$f(x') = \frac{1}{\lambda} \cdot (\exp(-x') - 1), \tag{3.2.9}$$

i.e., equivalently,
$$f(x') = \frac{1}{|\lambda|} \cdot (1 - \exp(-x')). \tag{3.2.10}$$

In both cases, we can conclude that a Sugeno λ-measure is indeed equivalent to a probability measure.

First comment: how unique is the transformation $f(x)$? If we have two different functions $f(x)$ and $f'(x)$ with the above property, then for each triple (a, b, c) that satisfies the formula (3.2.3), we will have $c' = a' + b'$ and $c'' = a'' + b''$, where $a'' = (f')^{-1}(a)$, $b'' = (f')^{-1}(b)$, and $c'' = (f')^{-1}(c)$. Thus, a mapping $x' = h(x'')$, where $h(x) \stackrel{\text{def}}{=} f^{-1}(f'(x))$, has the property that $c'' = a'' + b''$ implies $h(c'') = h(a'') + h(b'')$.

It is known that the only mappings from non-negative numbers to non-negative numbers that satisfy this property are linear functions $h(k) = k \cdot x$. Thus, once we know one such function $f(x)$, all other functions $f'(x)$ satisfy the property that

$f^{-1}(f'(x)) = k \cdot x$. By applying the function $f(x)$ to both sides of this equality, we conclude that

$$f'(x) = f(k \cdot x).$$

In other words, all such functions can be obtained from each other by an appropriate linear re-scaling $x \to k \cdot x$.

Second comment: why not also require that $f(1) = 1$. We are looking for a function $f(x)$ that transforms the probability $p(A)$ into a fuzzy measure $g(A) = f(p(A))$. For the functions (3.2.8) and (3.2.10), for $x = 0$, we have $f(0) = 0$. This equality is in good accordance with the fact that for $A = \emptyset$, we have $p(\emptyset) = g(\emptyset) = 0$ and thus, we should have $0 = g(\emptyset) = f(p(\emptyset)) = f(0)$.

Similarly, it makes sense to consider $A = X$; in this case, we have $p(X) = g(X) = 1$ and thus, we should have $1 = g(X) = f(p(X)) = f(1)$, i.e., $f(1) = 1$. Let us show that we can use the above non-uniqueness to satisfy this additional property. Indeed, once we have found the function $f(x)$, any function $f'(x) = f(k \cdot x)$ also has the desired property. We can therefore select k so that we will have the desired property $f'(1) = f(k \cdot 1) = 1$. This equality means $f(k) = 1$, so we should choose $k = f^{-1}(1)$.

Specifically, for $\lambda > 0$, we have $f^{-1}(x) = \ln(1 + \lambda \cdot x)$, so $k = \ln(1 + \lambda)$, and thus,

$$f'(x') = \frac{1}{\lambda} \cdot \left(\exp(\ln(1 + \lambda) \cdot x') - 1 \right) = \frac{1}{\lambda} \cdot \left((1 + \lambda)^{x'} - 1 \right).$$

Similarly, for $\lambda < 0$, we have $f^{-1}(x) = -\ln(1 - |\lambda| \cdot x)$, so $k = -\ln(1 - |\lambda|)$, and thus,

$$f'(x') = \frac{1}{|\lambda|} \cdot \left(1 - \exp(\ln(1 - |\lambda|) \cdot x') \right) = \frac{1}{|\lambda|} \cdot \left(1 - (1 - |\lambda|)^{x'} \right).$$

So why do we need Sugeno measures? Because of the equivalence, we can view the values of the Sugeno measure as simply re-scaling probabilities $g(A) = f(p(A))$ for the corresponding probability measure.

So why not just store the corresponding probability values $p(A)$? In other words, why not just re-scale all the values $g(A)$ into the corresponding probability values $p(A) = f^{-1}(g(A))$? At first glance, this would be a win-win arrangement, because once we do this re-scaling, we can simply use known probabilistic techniques.

What we plan to do. Let us show that:

- while from the purely *mathematical* viewpoint, a Sugeno λ-measure is equivalent to a probability measure,
- from the *computational* viewpoint, the direct use of Sugeno measures is much more efficient.

To explain this advantage, let us clarify what we mean by direct use of Sugeno measure and what we mean by an alternative of using a reduction to a probability measure.

The corresponding computational problem: a brief description. We are interested in understanding the degree of possibility of different sets of events. These degrees $g(A)$ come from an expert.

Theoretically, we could ask the expert to provide us with the values $g(A)$ corresponding to all possible sets A, but this would require an unrealistically large number of questions.

A feasible alternative is to elicit some values $g(A)$ from the experts and then use these values to estimate the missing values $g(A)$. A possibility of such estimation follows from the definition of a Sugeno λ-measure. Namely, once we know the values $g(A)$ and $g(B)$ corresponding to a two disjoint sets A and B, we do not need to additionally elicit, from this expert, the degree $g(A \cup B)$: this degree can be estimated based on the known values $g(A)$ and $g(B)$.

Let us explain how the desired degree $g(A \cup B)$ can be estimated.

Estimating $g(A \cup B)$ by a directly use of Sugeno measure. The first alternative is to simply estimate the degree $g(A \cup B)$ by applying the formula (3.2.2).

Estimating $g(A \cup B)$ by a reduction to a probability measure. An alternative idea—which is likely to be used by a practitioner accustomed to the probabilistic data processing—is to use the above-described reduction to a probability measure. Namely:

- first, we use the known reduction to find the corresponding values of the probabilities
$$p(A) = f^{-1}(g(A)) \text{ and } p(B) = f^{-1}(g(B));$$

- then, we add these probabilities to get
$$p(A \cup B) = p(A) + p(B);$$

- finally, we re-scale this resulting probability back into degree-of-confidence scale by applying the function $f(x)$ to this value $p(A \cup B)$, i.e., we compute
$$g(A \cup B) = f(p(A \cup B)).$$

Direct use of Sugeno measure is computationally more efficient. If we directly use Sugeno measure, then all we need to do is add and multiply. Inside a computer, both addition and multiplication are very directly hardware supported and therefore very fast.

In contrast, the use of reduction to probability measures requires that we compute the value of logarithm (to compute $f^{-1}(x)$) and exponential function (to compute $f(x)$). These computations are much slower than elementary arithmetic operations.

Thus, the direct use of Sugeno measure is definitely much more computationally efficient.

Comment. Of course, this result does not mean that we *should* always use Sugeno measures: for example, if we use the operation once and off-line, spending one microsecond more to compute exp and ln is not a big deal. However, in time-critical situations, with limited computational ability—e.g., if we are embedding some AI abilities into a control chip—saving computation time is important.

How to explain the use of Sugeno measure to a probabilist. The above argument enables us to explain the use of Sugeno measure to a person who is either skeptical about (or unfamiliar with) fuzzy measures. This explanation is as follows.

We are interested in expert estimates of probabilities of different sets of events. It is known that expert estimates of the probabilities are biased (see Chap. 1): the expert's subjective estimates $g(A)$ of the corresponding probabilities $p(A)$ are equal to $g(A) = f(p(A))$ for an appropriate re-scaling function $f(A)$.

In this case, a natural ideas seems to be:

- to re-scale all the estimates back into the probabilities, i.e., to estimate these probabilities $p(A)$ as

$$p(A) = f^{-1}(g(A)),$$

and then
- to use the usual algorithms to process these probabilities.

In particular, if we know the expert's estimates $g(A)$ and $g(B)$ corresponding to two disjoint sets A and B, and we want to predict the expert's estimate $g(A \cup B)$ corresponding to their union, then we:

- first, re-scale the values $g(A)$ and $g(B)$ into the un-biased probability scale, i.e., compute

$$p(A) = f^{-1}(g(A)) \text{ and } p(B) = f^{-1}(g(B));$$

- then, we compute

$$p(A \cup B) = p(A) + p(B);$$

- finally, we estimate $g(A \cup B)$ by applying the corresponding biasing function $f(x)$ to the resulting probability:

$$g(A \cup B) = f(p(A \cup B)).$$

It turns out that for some biasing functions $f(x)$, it is computationally more efficient *not* to re-scale into probabilities, but to store and process the original biased values $g(A)$. This is, in effect, the essence of applications of a Sugeno λ-measure are about.

Which other fuzzy measures have this property? We want to describe fuzzy measures which are mathematically equivalent to probability measures, but in which processing the fuzzy measure directly is more efficient than using the corresponding reduction to a probability measure.

Equivalence of a fuzzy measure $g(A)$ to a probability measure $p(A)$ means that $g(A) = f(p(A))$ for some 1-1 function $f(x)$. For the empty set $A = \emptyset$, for both measures, we should have zeros $p(\emptyset) = g(\emptyset) = 0$, so we should have $f(0) = 0$.

In general, for such fuzzy measures, if we know the values $g(A)$ and $g(B)$ corresponding to two disjoint sets A and B, then we can compute the measure $g(A \cup B)$ as follows:

- first, we use the known values $g(A)$ and $g(B)$ and the relations

$$g(A) = f(p(A)) \text{ and } g(B) = f(p(B))$$

to reconstruct the corresponding probability values

$$p(A) = f^{-1}(g(A)) \text{ and } p(B) = f^{-1}(g(B)),$$

where $f^{-1}(x)$ denotes the inverse function;
- then, we add the resulting values $p(A)$ and $p(B)$, and get

$$p(A \cup B) = p(A) + p(B);$$

- finally, we transform this probability back into the fuzzy measure, as

$$g(A \cup B) = f(p(A \cup B)).$$

The resulting value can be described by an explicit expression

$$g(A \cup B) = f(f^{-1}(g(A)) + f^{-1}(g(B))),$$

i.e., an expression of the type

$$g(A \cup B) = F(g(A), g(B)),$$

where

$$F(a, b) \stackrel{\text{def}}{=} f(f^{-1}(a) + f^{-1}(b)) \tag{3.2.11}$$

For $a = b = 0$, due to $f(0) = 0$, we get $F(0, 0) = 0$.

We are looking for situations in which the direct computation of a function $F(a, b)$ is more computationally efficient than using the above three-stage scheme. Thus, we are looking for situations in which the corresponding function $F(a, b)$ can be computed fast. In the computer, the fastest elementary operations are the hardware-supported ones: addition, subtraction, multiplication, and division. So, we should be looking for the functions $F(a, b)$ that can be computed by using only these four arithmetic operations.

In mathematical terms, functions that can be computed from the unknowns and constants by using only addition, subtraction, multiplication, and division are known as *rational functions* (they can be always represented as ratios of two polynomials). In these terms, we are looking for situations in which the corresponding aggregation function is rational.

Natural properties of the aggregation function. From the formula (3.2.11), we can easily conclude that the operation $F(a, b)$ is commutative: $F(a, b) = F(b, a)$.

One can also easily check that this operation is associative. Indeed, by (3.2.11), we have

$$F(F(a, b), c) = f(f^{-1}(F(a, b)) + f^{-1}(c)).$$

From (3.2.11), we conclude that

$$f^{-1}(F(a, b)) = f^{-1}(a) + f^{-1}(b).$$

Thus,

$$F(F(a, b), c) = f(f^{-1}(a) + f^{-1}(b) + f^{-1}(c)).$$

The right-hand side does not change if we change the order of the elements a, b, and c. Thus, we have

$$F(a, F(b, c)) = f(f^{-1}(a) + f^{-1}(b) + f^{-1}(c)),$$

i.e., $F(F(a, b), c) = F(a, F(b, c))$. So, the operation $F(a, b)$ is indeed associative.

Since we are looking for rational functions $F(a, b)$ for which $F(0, 0) = 0$, we are thus looking for rational commutative and associative operations $F(a, b)$ for which $F(0, 0) = 0$.

There is a known classification of all rational commutative associative binary operations. A classification of all possible rational commutative associative operations is known; it is described in [19]. Namely, the authors of [19] show that each such operation is "isomorphic" to either $x + y$ or $x + y + x \cdot y$, in the sense that there exists a fractional-linear transformation $a \to t(a)$ for which either $F(a, b) = t^{-1}(t(a) + t(b))$ or $F(a, b) = t^{-1}(t(a) + t(b) + t(a) \cdot t(b))$.

In other words, $F(a, b) = c$ means either than $t(c) = t(a) + t(b)$ or that $t(c) = t(a) + t(b) + t(a) \cdot t(b)$.

Comment. It should be mentioned that the paper [19] calls this relation by a fractional-linear transformation *equivalence*, not isomorphism; we changed the term since we already use the term "equivalence" in a different sense.

Let us use this known result. Let us use this result to classify the desired operations $F(a, b)$. First, we want an operation for which $F(0, b) = b$ for all b. In terms of t, this means that either $t(b) = t(b) + t(0)$ for all b, or $t(b) = t(b) + t(0) + t(0) \cdot t(b)$ for all b. In both cases, this implies that $t(0) = 0$. Thus, $t(a)$ is a fractional-linear function for which $t(0) = 0$.

A general fractional-linear function has the form

$$t(a) = \frac{p + q \cdot a}{r + s \cdot a}.$$

The fact that $t(0) = 0$ implies that $p = 0$, so we get

$$t(a) = \frac{q \cdot a}{r + s \cdot a}. \tag{3.2.12}$$

Here, we must have $r \neq 0$, because otherwise, the right-hand side of this expression is simply a constant q/s and not an invertible transformation. Since $r \neq 0$, we can divide both the numerator and the denominator of this expression by r and get a simplified formula

$$t(a) = \frac{A \cdot a}{1 + B \cdot a}, \tag{3.2.13}$$

where we denoted

$$A \overset{\text{def}}{=} \frac{q}{r} \text{ and } B \overset{\text{def}}{=} \frac{s}{r}.$$

For this transformation, the inverse transformation can be obtained from the fact that

$$a' = \frac{A \cdot a}{1 + B \cdot a}$$

implies

$$\frac{1}{a'} = \frac{1 + B \cdot a}{A \cdot a} = \frac{1}{A \cdot a} + \frac{B}{A}.$$

Thus,

$$\frac{1}{A \cdot a} = \frac{1}{a'} - \frac{B}{A},$$

so

$$A \cdot a = \frac{1}{\dfrac{1}{a'} - \dfrac{B}{A}}$$

and

$$a = \frac{\dfrac{1}{A}}{\dfrac{1}{a'} - \dfrac{B}{A}} = \frac{a'}{A - B \cdot a'}. \tag{3.2.14}$$

So, for operations equivalent to $x + y$, we get

$$c' = a' + b' = t(a) + t(b) = \frac{A \cdot a}{1 + B \cdot a} + \frac{A \cdot b}{1 + B \cdot b}.$$

Thus,

$$c = F(a, b) = t^{-1}(c') = \frac{\dfrac{A \cdot a}{1 + B \cdot a} + \dfrac{A \cdot a}{1 + B \cdot b}}{A - \dfrac{A \cdot B \cdot a}{1 + B \cdot a} - \dfrac{A \cdot B \cdot b}{1 + B \cdot b}}.$$

Dividing both numerator and denominator by the common factor A, we get

$$F(a, b) = \frac{\dfrac{a}{1 + B \cdot a} + \dfrac{b}{1 + B \cdot b}}{1 - \dfrac{B \cdot a}{1 + B \cdot a} - \dfrac{B \cdot b}{1 + B \cdot b}}. \tag{3.2.15}$$

Bringing the sums in the numerator and in the denominator to the common denominator and taking into account that this common denominator is the same for numerator and denominator of the expression (3.2.15), we conclude that

$$F(a, b) = \frac{a \cdot (1 + B \cdot b) + b \cdot (1 + B \cdot a)}{(1 + B \cdot a) \cdot (1 + B \cdot b) - B \cdot a - B \cdot b} =$$

$$\frac{a + b + 2B \cdot a \cdot b}{1 + B \cdot a + B \cdot b + B^2 \cdot a \cdot b - B \cdot a - B \cdot b}.$$

Finally, by cancelling equal terms in the denominator, we get the final formula

$$F(a, b) = \frac{a + b + 2B \cdot a \cdot b}{1 + B^2 \cdot a \cdot b}. \tag{3.2.16}$$

For operations equivalent to $x + y + x \cdot y$, we similarly get

$$c' = a' + b' + a' \cdot b' = t(a) + t(b) + t(a) \cdot t(b) = \frac{A \cdot a}{1 + B \cdot a} + \frac{A \cdot b}{1 + B \cdot b} + \frac{A^2 \cdot a \cdot b}{(1 + B \cdot a) \cdot (1 + B \cdot b)}.$$

If we bring these terms to a common denominator, we get

$$c' = \frac{A \cdot a \cdot (1 + B \cdot b) + A \cdot b \cdot (1 + B \cdot a) + A^2 \cdot x \cdot y}{(1 + B \cdot a) \cdot (1 + B \cdot b)} = \frac{A \cdot (a + b + (2B + A) \cdot a \cdot b)}{(1 + B \cdot a) \cdot (1 + B \cdot b)}.$$

Therefore,

$$F(a, b) = t^{-1}(c') = \frac{\dfrac{A \cdot (a + b + (2B + A) \cdot a \cdot b)}{(1 + B \cdot a) \cdot (1 + B \cdot b)}}{A - \dfrac{A \cdot B \cdot (a + b + (2B + A) \cdot a \cdot b)}{(1 + B \cdot a) \cdot (1 + B \cdot b)}}.$$

Dividing both the numerator and the denominator of this expression by A, we conclude that

$$F(a, b) = \frac{\dfrac{a + b + (2B + A) \cdot a \cdot b}{(1 + B \cdot a) \cdot (1 + B \cdot b)}}{1 - \dfrac{B \cdot (a + b + (2B + A) \cdot a \cdot b)}{(1 + B \cdot a) \cdot (1 + B \cdot b)}}.$$

By bringing the difference in the denominator to the common denominator, we get

$$F(a, b) = \frac{N}{D},$$

where

$$N \overset{\text{def}}{=} a + b + (2B + A) \cdot a \cdot b$$

and

$$D \overset{\text{def}}{=} 1 + B \cdot a + B \cdot b + B^2 \cdot a \cdot b - B \cdot a - B \cdot b - B \cdot (2B + A) \cdot a \cdot b =$$

$$1 - B \cdot (B + A) \cdot a \cdot b.$$

Thus

$$F(a, b) = \frac{a + b + (2B + A) \cdot a \cdot b}{1 - B \cdot (B + A) \cdot a \cdot b}. \tag{3.2.17}$$

Comment. Similarly to the case of Sugeno measure, we can always impose an additional requirement $f(1) = 1$, by replacing the original re-scaling $f(x)$ with a modified re-scaling $f'(x) \overset{\text{def}}{=} f(k \cdot x)$ with $k = f^{-1}(1)$ (for which $f(k) = 1$).

Summary. We consider the fuzzy measures $g(A)$ which are equivalent to probability measures. For such fuzzy measures, once we know the values $g(A)$ and $g(B)$ for two disjoint sets A and B, we can compute the degree $d(A \cup B)$ as $d(A \cup B) = F(g(A), g(B))$ for the corresponding aggregation operation $F(a, b)$.

This value $g(A \cup B)$ can be computed in two different ways:

- we can reduce the problem to the probability measures, i.e., compute the corresponding probabilities, add them up, and use this sum $p(A) + p(B)$ to compute the desired value $g(A \cup B)$;
- alternatively, we can compute the value $g(A \cup B)$ directly, as $F(g(A), g(B))$.

We are looking for operations for which the direct use of fuzzy measures is computationally faster, i.e., in precise terms, for which the aggregation operation can be computed by using fast (hardware supported) elementary arithmetic operations. It turns out that the only such operations are operations (3.2.16) and (3.2.17) corresponding to different values of A and B.

By using these operations, we thus get a class of fuzzy measures that naturally generalizes Sugeno λ-measures. Let us hope that fuzzy measures from this class will be as practically successful as Sugeno λ-measures themselves.

When do the original Sugeno λ-measures lie in this class? To understand it, let us recall that not all arithmetic operations require the same computation time. Indeed, addition is the simplest operation. Multiplication is, in effect, several additions, so multiplication take somewhat longer. Division requires several iterations, so it takes the longest time. So, any computation that does not include division is much faster. Of our formulas (3.2.16) and (3.2.17), the only cases when we do not use division are cases when $B = 0$, i.e., cases when we have $F(a, b) = a + b$ (corresponding to probability measures) and $F(a, b) = a + b + A \cdot a \cdot b$ corresponding to Sugeno λ-measures. From this viewpoint, Sugeno λ-measures are the ones for which the direct use of the fuzzy measure has the largest computational advantage over the reduction to probability measures.

Comment. Our result is similar to the known result that the only rational t-norms and t-conorms are Hamacher operations; see, e.g., [3, 20, 64]. The difference is in our analysis, we do not assume that the aggregation operation corresponding to the fuzzy measure is a t-conorm: for example for the Sugeno aggregation operation, $F(1, 1) = 1 + 1 + \lambda = 2 + \lambda > 1$, while for t-conorm, we always have $F(1, 1) = 1$.

Also, the result from [19] that we use here does not depend on the use of real numbers, it is true for any field—e.g., for subfields of the field of real numbers (such as the field of rational numbers) or for super-fields (such as fields that contain infinitesimal elements).

3.3 Making a Decision Under Uncertainty: A Possible Explanation of the Usual Safety Factor of 2

A usual decision making procedure assumes that we have complete information about the situation. In practice, in addition to *known* factors, however, there are usually *unknown* factors which also contribute to the uncertainty. To make proper decisions, we need to also take these "unknown unknowns" into account. In practice, this is usually done by multiplying the original estimate for inaccuracy by a "safety" factor of 2. In this section, we provide an explanation for this empirical factor.

Results from this section first appeared in [83].

What is a safety factor? When engineers design an accurate measuring instrument, they try their best to make it as accurate as possible. For this purpose, they apply known techniques to eliminate (or at least to drastically decrease) all major sources of error. For example:

- thermal noise can be drastically decreased by cooling,
- the effect of the outside electromagnetic fields can be drastically decreased by barriers made of conductive or magnetic materials,
- the effect of vibration can be decreased by an appropriate suspension, etc.

Once the largest sources of inaccuracy are (largely) eliminated, we need to deal with the next largest sources, etc. No matter how many sources of inaccuracy we deal with, there is always the next one. As a result, we can never fully eliminate all possible sources of inaccuracy.

At each stage of the instrument design, we usually understand reasonably well what is the main source of the remaining inaccuracy, and how to gauge the corresponding inaccuracy. As a result, we have a good estimate for the largest possible value Δ of the inaccuracy caused by this source.

This value Δ, however, does not provide a full description of the measurement inaccuracy. Indeed, in addition to the "known unknown" (the known main source of remaining inaccuracy), there are also "unknown unknowns"—smaller additional factors that also contribute to measurement inaccuracy.

To take these "unknown unknowns" into account, practitioners usually multiply Δ by a larger-than-one factor. This factor is known as the *safety factor*.

A related idea is used when we design engineering objects such as roads, bridges, houses, etc. For example, when we design a bridge, we want to make sure that its deformation caused by different loads does not exceed the limit after which catastrophic changes may occur. For this purpose, we first estimate the deformation caused by the known factors, and then—to be on the safe side—multiply this deformation by a safety factor.

Safety factor of 2 is most frequently used. In civil engineering, a usual recommendation is to use the safety factor of 2; see, e.g., [156].

This value has been in use for more than 150 years: according to [150], the earliest recorded use of this value can traced to a book published in 1858 [135].

This value is also widely used beyond standard civil engineering projects: e.g., it was used in the design of Buran, a successful Soviet-made fully automatic pilotless Space Shuttle [99].

Comment. It should be mentioned that in situations requiring extreme caution—e.g., in piloted space flights—usually, a larger value of safety factor is used, to provide additional safety.

Open problem. The fact that the safety factor of 2 has been in use for more than 150 years shows that this value is reasonable. However, there seems to be no convincing explanation of why this particular value is empirically reasonable.

In this section, we provide a possible theoretical explanation for this value.

Analysis of the problem. We know that in addition to the largest inaccuracy of size Δ, there is also next largest inaccuracy of size $\Delta_1 < \Delta$. Once we take that inaccuracy into account, then we will need to take into account the next inaccuracy, of size $\Delta_2 < \Delta_1$, etc. The final inaccuracy can be estimated as the sum of all these inaccuracies, i.e., as the sum $\Delta + \Delta_1 + \Delta_2 + \cdots$, where

$$\cdots < \Delta_{k+1} < \Delta_k < \cdots < \Delta_2 < \Delta_1 < \Delta.$$

To estimate this sum, it is therefore necessary to estimate all the sizes $\Delta_1, \Delta_2, \ldots$

Let us estimate these sizes one by one.

Estimating Δ_1. The only information that we have about the value Δ_1 is that it is larger than 0 and smaller than Δ. In other words, in principle, Δ_1 can take any value from the interval $(0, \Delta)$. We have no information about the probabilities of different values from this interval.

Since we have no reason to think that some specific value from this interval is more probable and some other specific value is less probable, it is reasonable to assume that all the values from this interval are equally probable. This argument—known as *Laplace principle of indifference*—is widely used in statistics applications; see, e.g., [59, 144]

In precise terms, this means that we assume that the distribution of possible values Δ_1 on the interval $(0, \Delta)$ is *uniform*. For the uniform distribution on an interval, the expected value is the interval's midpoint. Thus, the expected value of Δ_1 is equal to $\Delta/2$.

This expected value is what we will use as an estimate for Δ_1.

Estimating Δ_2, **etc**. Once we produced an estimate $\Delta_1 = \Delta/2$, a next step is to estimate the next inaccuracy component Δ_2. The only information that we have about the value Δ_2 is that it is larger than 0 and smaller than Δ_1. In other words, in principle, Δ_2 can take any value from the interval $(0, \Delta_1)$. We have no information about the probability of different values from this interval. Thus, similarly to the previous section, it is reasonable to assume that Δ_2 is uniformly distributed on the interval $(0, \Delta_1)$. In this case, the expected value of Δ_2 is equal to $\Delta_1/2$. This expected value $\Delta_2 = \Delta_1/2$ is what we will use as an estimate for Δ_2.

Similarly, we conclude that a reasonable estimate for $\Delta_3 < \Delta_2$ is $\Delta_3 = \Delta_2/2$, and, in general, that for every k, a reasonable estimate for $\Delta_{k+1} < \Delta_k$ is $\Delta_{k+1} = \Delta_k/2$.

Adding all these estimates leads to the desired explanation. From $\Delta_1 = \Delta/2$ and $\Delta_{k+1} = \Delta_k/2$, we can conclude, by induction, that $\Delta_k = 2^{-k} \cdot \Delta$. Substituting these estimates into the formula for the overall inaccuracy $\delta = \Delta + \Delta_1 + \Delta_2 + \cdots$, we conclude that

$$\delta = \Delta + 2^{-1} \cdot \Delta + 2^{-2} \cdot \Delta + \cdots = \left(1 + 2^{-1} + 2^{-2} + \cdots\right) \cdot \Delta.$$

The sum of the geometric progression $1 + 2^{-1} + 2^{-2} + \cdots$ is 2, so we get $\delta = 2\Delta$.

This is exactly the formula that we tried to explain—with a safety factor of two.

3.4 Taking into Account Interests of Others: Why Awe Makes People More Generous

In the previous sections, we only took into account the decision maker's own interests. In practice, we also need to take into account interests of others. This is a well-studied topic in decision making, there have many experiments explaining how people perform such group decision making, and there are reasonable explanations for most of these experiments. However, there are some recent experiments for which no reasonable explanation has been proposed yet. In this section, we provide an explanation of these experimental observations.

The results of this paper first appeared in [91].

Recent experiment: a brief summary. A recent experiment [129] showed that the feeling of awe increases people's generosity.

Recent experiment: details. To measure a person's generosity, researchers use a so-called *Ultimatum Bargaining Game*. In this game, to test a person's generosity, this person is paired with another person in a simulated situation in which they should share a certain fixed amount of money given to them by the experimenter:

- the tested person announces which part of this amount he or she is willing to give to the second person;
- if the second person agrees, each of them gets the agreed-on amount;
- if the offered part is too small and the second person disagrees, no one gets anything.

The generosity is then measured by the amount of money that the tested person is willing to give to his/her companion.

The gist of the experiment is that this amount increases when a tested person has a feeling of awe, which is induced:

- either by requiring the persons to think about awe-inspiring natural scenes,
- or by explicitly showing such scenes prior to testing generosity.

Comment. It is worth mentioning that a similar increase in the transferred amount of money was observed in a different situation, when the tested person was injected with euphoria-inducing oxytocin [66].

How the results of this experiment are currently explained. The paper [129] provides the following qualitative explanation of this result: that the presence of awe leads to a feeling of smaller self. This, in turn, makes the person more social and thus, more generous.

It is desirable to have a more quantitative explanation. The above qualitative explanation is reasonable, but it is not necessarily fully convincing: the feeling of awe caused by a magnificent nature scene definitely decreases the feeling of importance of self—but it also decreases the feeling of importance of other people. It would make perfect sense if the feeling of awe led to more donations to nature conservation funds, but why to other people?

What we plan to do. In order to come up with a more convincing explanation of the above experiment, we analyze this experiment in quantitative terms, by using the standard utility-based approach to decision making; see, e.g., [35, 97, 114, 133]. Our analysis shows that indeed, within this approach, the presence of awe leads to an increase in generosity.

The notion of utility: reminder. As we have mentioned earlier, according to decision theory, a rational person's preferences can be described by his/her *utility function*, a function that assigns, to each alternative, a real number in such a way that out of several alternatives, the person always selects the one whose utility is the largest.

How utility depends on the amount of money. Experiments have shown that for situations with monetary gain, utility u grows with the money amount m as $u \approx m^\alpha$, with $\alpha \approx 0.5$, i.e., approximately as $u \approx \sqrt{m}$; see, e.g., [61] and references therein.

So, if the given amount of money m is distributed between two participants, so that the first person gets m_1 and the second person gets $m_2 = m - m_1$, then:

- the utility of the first person is $u_1(m_1) = \sqrt{m_1}$ and
- the utility of the second person is $u_2(m_1) = \sqrt{m_2} = \sqrt{m - m_1}$.

Comment. The specific dependence $u \approx \sqrt{m}$ can itself be explained by utility-based decision theory (see above).

The dependence of utility on other "units of pleasure". It is reasonable to assume that the formula $u \approx \sqrt{m}$ describes not only the dependence of utility on the amount of money, but also on the overall amount of "units of pleasure" received by a person, be it money, material good, or feeling of awe.

So, if we denote, by a, the amount of such units corresponding to the feeling of awe, then, if the first person also gets the amount of money m_1, the overall amount of such units is $a + m_1$, and thus, this persons' utility is approximately equal to $u_1(m_1) = \sqrt{a + m_1}$.

By definition, awe means that the corresponding pleasure is much larger than what one normally gets from a modest money amount, i.e., that $a \gg m_1$.

Effect of empathy. A person's preferences depend not only on what this person gets, they also depend on what others get. Normally, this dependence is positive, i.e., we feel happier if other people are happy.

The idea that a utility of a person depends on utilities of others was first described in [136, 137]. It was further developed by another future Nobelist Gary Becker; see, e.g., [9]; see also [18, 38, 55, 114, 154].

If we take empathy into account, then, instead of the original no-empathy values $\sqrt{m_1}$ and $\sqrt{m - m_1}$, we get values

$$u_1(m_1) = \sqrt{m_1} + \alpha_{12} \cdot \sqrt{m - m_1}$$

and

$$u_2(m_1) = \sqrt{m - m_1} + \alpha_{21} \cdot \sqrt{m_1},$$

where $\alpha_{ij} > 0$ are positive numbers. People participating the above experiment are strangers to each other, so their mutual empathy is not large: $\alpha_{ij} \ll 1$.

In the presence of awe, we similarly get $u_1(m_1) = \sqrt{a + m_1} + \alpha_{12} \cdot \sqrt{m - m_1}$ and $u_2(m_1) = \sqrt{m - m_1} + \alpha_{21} \cdot \sqrt{a + m_1}$.

How joint decisions are made. In the Ultimatum Bargaining Game, two participants need to cooperate to get money. For such cooperative situations, an optimal solution has been discovered by the Nobelist John Nash [97, 112, 114]: a group should select the alternative x for which the following product attains its largest possible value:

$$(u_1(x) - u_1(0)) \cdot (u_2(x) - u_2(0)),$$

where $u_i(x)$ is the i-th person utility corresponding to the alternative x and $u_i(0)$ is this person's utility in the original (*status quo*) situation. This is known as *Nash's bargaining solution*.

The fact that Nash's bargaining solution can be used to describe such games is emphasized, e.g., in [67]. In our case, the status quo situation is when neither the first nor the second participant get any money, i.e., when $m_1 = 0$ and $m_2 = 0$. In the absence of a, this means $u_1(0) = u_2(0) = 0$. In the presence of awe, this means that:

- in the first approximation, when we ignore empathy, we get $u_1(0) = \sqrt{a}$ and $u_2(0) = 0$;
- when we take empathy into account, we get $u_1(0) = \sqrt{a}$ and $u_2(0) = \alpha_{21} \cdot \sqrt{a}$.

Resulting formulation of the problem. Now, we are ready to formulate the situation in precise terms. We compare the optimal amounts $m_2 = m - m_1$ corresponding to two different situations.

In the first situation, there is no awe, so we select the value m_1 for which the following product attains the largest possible value:

$$\left(\sqrt{m_1} + \alpha_{12} \cdot \sqrt{m - m_1}\right) \cdot \left(\sqrt{m - m_1} + \alpha_{21} \cdot \sqrt{m_1}\right). \tag{3.4.1}$$

In the second situation, there is an awe $a \gg m_1$, so we select the value a for which the following product attains the largest possible value:

$$\left(\sqrt{a + m_1} - \sqrt{a} + \alpha_{12} \cdot \sqrt{m - m_1}\right) \cdot \left(\sqrt{m - m_1} + \alpha_{21} \cdot \left(\sqrt{a + m_1} - \sqrt{a}\right)\right). \tag{3.4.2}$$

Let us estimate and compare these optimal values.

Analysis of the first (no-awe) situation. Let us start with analyzing the first situation. Since the values α_{ij} are small, in the first approximation, we can safely ignore the corresponding terms and instead maximize the simplified product $\sqrt{m_1} \cdot \sqrt{m - m_1}$.

Maximizing this product is equivalent to maximizing its square, i.e., the value $m_1 \cdot (m - m_1)$. Differentiating this expression and equating the derivative to 0, we conclude that the maximum is attained when $m_1 = 0.5 \cdot m$ and $m_2 = m - m_1 = 0.5 \cdot m$. This is indeed close to the observed division in the Ultimatum Bargaining Game [67, 129].

Analysis of the second (awe) situation. In the second situation, we can use another simplifying approximation: namely, since $a \gg x_1$, we can use the fact that in general, for a differentiable function $f(x) = \sqrt{x}$, we have $\dfrac{df}{dx} = \lim_{h \to 0} \dfrac{f(x+h) - f(x)}{h}$.

Thus, for small h, we have $\dfrac{df}{dx} \approx \dfrac{f(x+h) - f(x)}{h}$, hence $f(x+h) - f(x) \approx \dfrac{df}{dx} \cdot h$.

In particular, for $f(x) = \sqrt{x}$, we get $\sqrt{a + m_1} - \sqrt{a} \approx \dfrac{1}{2\sqrt{a}} \cdot m_1$. Since the value

a is huge, the ratio $\dfrac{1}{2\sqrt{a}}$ is very small, so, in the first approximation, we can safely ignore this ratio in comparison with the term $\alpha_{12} \cdot \sqrt{m - m_1}$. Similarly, in the second factor, we can safely ignore the term $\alpha_{21} \cdot \left(\sqrt{a + m_1} - \sqrt{a} \right)$ which is proportional to this ratio. Thus, in this first approximation, maximization of the product (3.4.2) can be reduced to maximizing the following simplified product:

$$\alpha_{12} \cdot \sqrt{m - m_1} \cdot \sqrt{m - m_1} = \alpha_{12} \cdot (m - m_1).$$

Among all possible values m_1 from the interval $[0, m]$, the largest value of this expression is attained when $m_1 = 0$ and $m_2 = m - m_1 = m$, i.e., which indeed corresponds to the maximum generosity.

Conclusion: awe does increase generosity. As we have mentioned earlier, generosity is hereby measured by the amount of money $m_2 = m - m_1$ given to the second person.

We have shown that in the first approximation:

- in the first (no-awe) situation, the amount m_2 given to the second person is $m_2 = 0.5 \cdot m$, while
- in the second (awe) situation, the amount m_2 given to the second person is $m_2 = m$,

In this first approximation, since $m > 0.5 \cdot m$, the presence of awe does increase generosity.

Of course, there are solutions to the approximate problems, and thus, approximations to the solutions to the actual optimization problems. For the actual optimal solutions, we will have $m_2 \approx 0.5 \cdot m$ in the no-awe case and $m_2 \approx m$ in the awe case. Thus, still, the generosity in the awe case is larger. So, the utility-based decision theory indeed explains why awe increases generosity.

Comment. Since oxytocin also brings a large amount a of positive emotions, this model can also explain the above-mentioned results from [66].

3.5 Education as an Example of Joint Decision Making

In the previous section, we considered situations when it is important to take into account utility of others. An important extreme example of such process is *education*, when the main objective of the *teacher* is to increase the utility of the *student*, by making the student more knowledgeable. In this section, we will consider decision making problems related to education.

Specifically, in Sect. 3.5.1, we will consider the "how" of education, i.e., decisions related to education. In Sect. 3.5.2, we will analyze the results of these decisions.

3.5.1   The "How" of Education: Coming up with a Good Question Is Not Easy

The ability to ask good questions is an important part of learning skills. Coming up with a good question, a question that can really improve one's understanding of the topic, is not easy. In this subsection, we prove—on the example of probabilistic and fuzzy uncertainty—that the problem of selecting a good question is indeed hard.

The results from this subsection first appeared in [94].

Asking good questions is important. Even after a very good lecture, some parts of the material remain not perfectly clear. A natural way to clarify these parts is to ask questions to the lecturer.

Ideally, we should be able to ask a question that immediately clarifies the desired part of the material. Coming up with such good questions is an important part of learning process, it is a skill that takes a long time to master.

Coming up with good questions is not easy: an empirical fact. Even for experienced people, it is not easy to come up with a good question, i.e., with a question that will maximally decrease uncertainty.

What we do in this subsection. In this subsection, we prove that the problem of designing a good question is indeed computationally difficult (NP-hard).

We will show this both for probabilistic and for fuzzy uncertainty. Specifically, we will prove NP-hardness for the simplest types of questions—for "yes"-"no" questions for which the answer is "yes" or "no". Since already designing such simple questions is NP-hard, any more general problem (allowing more complex problems) is NP-hard as well.

What is uncertainty: a general description. A complete knowledge about any area—be it a physical system or an algorithm—would mean that we have the full description of the corresponding objects. From this viewpoint, uncertainty means that several different variants are consistent with our (partial) knowledge, and we are not sure which of these variants is true.

In the following text, we will denote possible variants by $v_1, v_2, \ldots, v_n$, or, if this does not cause any ambiguity, simply by $1, 2, \ldots, n$.

How to describe a "yes"-"no" question in these terms. A "yes"-"no" question is a question, an answer to which eliminates possible variants. In general, this means that after we get the answer to this question, instead of the original set $\{1, \ldots, n\}$ of possible variants, we have a smaller set:

- if the answer is "yes", then we are limited to the set $Y \subset \{1, \ldots, n\}$ of all the variants which are consistent with the "yes"-answer;
- if the answer is "no", then we are limited to the set $N \subset \{1, \ldots, n\}$ of all the variants which are consistent with the "no"-answer.

These two sets are complements to each other.

Examples. In some cases, we are almost certain about a certain variant, i.e., variant v_1. In this case, a natural question to ask if whether this understanding is correct. For this question:

- the "yes"-set Y consists of the single variant v_1, while
- the "no"-set $\{2, \ldots, n\}$ contains all other variants.

In other cases, we are completely unclear about the topic, e.g., we are completely unclear what is the numerical value of a certain quantity, we are not even sure whether this value is positive or non-positive. In this case, a natural question is to ask whether the actual value is positive. In such situation, each of the sets Y and N contain approximately a half of all original variants.

Probabilistic approach to describing uncertainty: a description. In the probabilistic approach, we assign a probability $p_i \geq 0$ to each of the possible variants, so that these probabilities add up to 1: $\sum_{i=1}^{n} p_i = 1$. The probability p_i, for example, may describe the frequency with which the i-th variant turned out to be true in similar previous situations.

How to quantify an amount of uncertainty: probabilistic case. In the case of probabilistic uncertainty, there is a well-established way to gauge the amount of uncertainty: namely, the entropy [46, 162]

$$S = -\sum_{i=1}^{n} p_i \cdot \ln(p_i). \tag{3.5.1}$$

This is a good estimate for the amount that we want to decrease by asking an appropriate question.

How do we select a question: idea. We would thus like to find the question that maximally decreases the uncertainty. Since in the probabilistic case, uncertainty is measured by entropy, we thus want to find a question that maximally decreases entropy.

How the answer changes the entropy. Once we know the answer to our "yes"-"no" question, the probabilities change.

If the answer was "yes", this means that the variants from the "no"-set N are no longer possible. For such variants $i \in N$, the new probabilities are 0s: $p_i' = 0$. For variants from the "yes"-set Y, the new probability is the conditional probability under the condition that the variant is in the "yes"-set, i.e.,

$$p_i' = p(i \mid Y) = \frac{p_i}{p(Y)}, \tag{3.5.2}$$

where the probability $p(Y)$ of the "yes"-answer is equal to the sum of the probabilities of all the variants that lead to the "yes"-answer:

$$p(Y) = \sum_{i \in Y} p_i. \tag{3.5.3}$$

Based on these new probabilities, we can compute the new entropy value

$$S' = -\sum_{i \in Y} p_i' \cdot \ln(p_i'). \tag{3.5.4}$$

On the other hand, if the answer was "no", this means that the variants from the "yes"-set Y are no longer possible. For such variants $i \in Y$, the new probabilities are 0s: $p_i'' = 0$. For variants from the "no"-set N, the new probability is the conditional probability under the condition that the variant is in the "no"-set, i.e.,

$$p_i'' = p(i \mid N) = \frac{p_i}{p(N)}, \tag{3.5.5}$$

where the probability $p(N)$ of the "no"-answer is equal to the sum of the probabilities of all the variants that lead to the "no"-answer:

$$p(N) = \sum_{i \in N} p_i. \tag{3.5.6}$$

Based on these new probabilities, we can compute the new entropy value

$$S'' = -\sum_{i \in N} p_i'' \cdot \ln(p_i''). \tag{3.5.7}$$

In the case of the "yes" answer, the entropy decreases by the amount $S - S'$. In the case of the "no"-answer, the entropy decreases by the amount $S - S''$. We know the probability $p(Y)$ of the "yes"-answer and we know the probability $p(N)$ of the "no"-answer. Thus, we can estimate the expected decrease in uncertainty as

$$\overline{S}(Y) = p(Y) \cdot (S - S') + p(N) \cdot (S - S''). \tag{3.5.8}$$

Thus, we arrive at the following formulation of the problem in precise terms.

Formulation of the problem in precise terms.

- We are *given* the probabilities $p_1, \ldots, p_n$ for which

$$\sum_{i=1}^{n} p_i = 1.$$

- We need to *find* a set $Y \subset \{1, \ldots, n\}$ for which the expected decrease in uncertainty $\overline{S}(Y)$ is the largest possible.

Here, $\overline{S}(Y)$ is described by the formula (3.5.8), and the components of this formula are described in formulas (3.5.1)–(3.5.7).

Formulation of the main probability-related result. Our main result is that the above problem—of coming up with the best possible question—is NP-hard.

What is NP-hard: a brief reminder. In many real-life problems, we are looking for a string (or for a sequence of a priori bounded numbers) that satisfies a certain property. For example, in the *subset sum* problem, we are given positive integers $s_1, \ldots, s_n$ representing the weights, and we need to divide these weights into two groups with exactly the same weight. In precise terms, we need to find a set $I \subseteq \{1, \ldots, n\}$ for which

$$\sum_{i \in I} s_i = \frac{1}{2} \cdot \left(\sum_{i=1}^{n} s_i \right).$$

The desired set I can be described as a sequence of n 0s and 1s, in which the i-th term is 1 if $i \in I$ and 0 if $i \notin I$.

In principle, we can solve each such problem by simply enumerating all possible strings, all possible combinations of numbers, etc. For example, in the above case, we can try all 2^n possible subsets of the set $\{1, \ldots, n\}$; this way, if there is a set I with the desired property, we will find it. The problem with this approach is that for large n, the corresponding number 2^n of computational steps becomes unreasonably large. For example, for $n = 300$, the resulting computation time exceeds lifetime of the Universe.

So, a natural question is: when can we solve such problems in feasible time, i.e., in time that does not exceed a polynomial of the size of the input? It is not known whether all exhaustive-search problems can be thus solved—this is the famous $P \overset{?}{=} NP$ problem. Most computer science researchers believe that some exhaustive-search problems cannot be feasibly solved—but in general, this remains an open problem.

What is known is that some problems are the hardest (NP-hard) in the sense that any exhaustive-search problem can be feasibly reduced to this problem. This means that, unless all exhaustive-search problems can be feasibly solved (which most computer scientists believe to be impossible), this particular problem cannot be feasibly solved.

The above subset sum problem has been proven to be NP-hard, as well as many other similar problems; see, e.g., [124].

How can we prove NP-hardness. As we have mentioned, a problem is NP-hard if every other exhaustive-search problem $\mathscr{Q}$ can be reduced to it. So, if we know that a problem $\mathscr{P}_0$ is NP-hard, then every $\mathscr{Q}$ can be reduced to it. Thus, if $\mathscr{P}_0$ can be reduced to our problem $\mathscr{P}$, then, by transitivity, any problem $\mathscr{Q}$ can be reduced to $\mathscr{P}$, i.e., $\mathscr{P}$ is indeed NP-hard.

Thus, to prove that a given problem is NP-hard, it is sufficient to reduce one known NP-hard problem $\mathscr{P}_0$ to this problem $\mathscr{P}$.

What we will do. To prove that the problem $\mathscr{P}$ of selecting a good question is NP-hard, we will follow the above idea. Namely, we will prove that the subset sum problem $\mathscr{P}_0$ (which is known to be NP-hard) can be reduced to $\mathscr{P}$.

Let us simplify the expression for $\overline{S}(Y)$. To build the desired reduction, let us simplify the expression (3.5.8). This expression uses the entropies S' and S''. So, to get the desired simplification, we will start with simplifying the expressions (3.5.4) and (3.5.7) for S' and S''.

Simplifying the expression for S'. Substituting the expression (3.5.2) for p_i' into the formula (3.5.4), we get

$$S' = -\sum_{i \in Y} \frac{p_i}{p(Y)} \cdot \ln\left(\frac{p_i}{p(Y)}\right).$$

All the terms in this sum are divided by $p(Y)$, so we can move this common denominator outside the sum:

$$S' = -\frac{1}{p(Y)} \cdot \left(\sum_{i \in Y} p_i \cdot \ln\left(\frac{p_i}{p(Y)}\right)\right).$$

The logarithm of the ratio is equal to the difference of logarithms, so we get

$$S' = -\frac{1}{p(Y)} \cdot \left(\sum_{i \in Y} p_i \cdot (\ln(p_i) - \ln(p(Y)))\right).$$

We can separate the terms proportional to $\ln(p_i)$ and to $\ln(p(Y))$ into two different sums. As a result, we get

$$S' = -\frac{1}{p(Y)} \cdot \left(\sum_{i \in Y} p_i \cdot \ln(p_i)\right) + \frac{1}{p(Y)} \cdot \left(\sum_{i \in Y} p_i \cdot \ln(p(Y))\right). \tag{3.5.9}$$

In the second sum, the factor $\ln(p(Y))$ does not depend on i and can, thus, be moved out of the summation:

$$\sum_{i \in Y} p_i \cdot \ln(p(Y)) = \ln(p(Y)) \cdot \sum_{i \in Y} p_i.$$

Here, the sum $\sum_{i \in Y} p_i$ is simply equal to $p(Y)$, so

$$\sum_{i \in Y} p_i \cdot \ln(p(Y)) = \ln(p(Y)) \cdot p(Y). \tag{3.5.10}$$

Substituting the expression (3.5.10) into the formula (3.5.9), and cancelling the terms $p(Y)$ in the numerator and in the denominator, we conclude that

$$S' = -\frac{1}{p(Y)} \cdot \left(\sum_{i \in Y} p_i \cdot \ln(p_i) \right) + \ln(p(Y)). \tag{3.5.11}$$

Simplifying the expression for S''. Similarly, we get

$$S'' = -\frac{1}{p(N)} \cdot \left(\sum_{i \in N} p_i \cdot \ln(p_i) \right) + \ln(p(N)). \tag{3.5.12}$$

Resulting simplification of the expression for $\overline{S}(Y)$. Since $p(Y) + p(N) = 1$, the expression (3.5.8) for $\overline{S}(Y)$ can be alternatively described as

$$\overline{S}(Y) = S - (p(Y) \cdot S' + p(N) \cdot S''). \tag{3.5.13}$$

By using expressions (3.5.11) and (3.5.12) for S' and S'', we conclude that

$$p(Y) \cdot S' + p(N) \cdot S'' =$$

$$-\frac{p(Y)}{p(Y)} \cdot \left(\sum_{i \in Y} p_i \cdot \ln(p_i) \right) + p(Y) \cdot \ln(p(Y)) - \frac{p(N)}{p(N)} \cdot \left(\sum_{i \in N} p_i \cdot \ln(p_i) \right) + p(N) \cdot \ln(p(N)) =$$

$$-\sum_{i \in Y} p_i \cdot \ln(p_i) - \sum_{i \in N} p_i \cdot \ln(p_i) + p(Y) \cdot \ln(p(Y)) + p(N) \cdot \ln(p(N)). \tag{3.5.14}$$

Since Y and N are complements to each other, we have

$$-\sum_{i \in Y} p_i \cdot \ln(p_i) - \sum_{i \in N} p_i \cdot \ln(p_i) = -\sum_{i=1}^{n} p_i \cdot \ln(p_i) = S.$$

Thus, the formula (3.5.14) takes the form

$$p(Y) \cdot S' + p(N) \cdot S'' = S + p(Y) \cdot \ln(p(Y)) + p(N) \cdot \ln(p(N)).$$

Therefore, the expression (3.5.13) takes the form

$$\overline{S}(Y) = -p(Y) \cdot \ln(p(Y)) - p(N) \cdot \ln(p(N)). \tag{3.5.15}$$

Here, $p(N) = 1 - p(Y)$, we have

$$\overline{S}(Y) = -p(Y) \cdot \ln(p(Y)) - (1 - p(Y)) \cdot \ln(1 - p(Y)). \qquad (3.5.16)$$

Resulting reduction. We want to find the set Y that maximizes the expected decrease in uncertainty, i.e., that maximizes the expression (3.5.16). Thus, we need to select a question Y for which $p(Y) = 0.5$.

Let us show that a subset sum problem can be reduced to this problem. Indeed, let us assume that we are given n positive integers $s_1, \ldots, s_n$. Then, we can form n probabilities

$$p_i \stackrel{\text{def}}{=} \frac{s_i}{\displaystyle\sum_{j=1}^{n} s_j} \qquad (3.5.17)$$

that add to 1. For this problem, if we can find a set Y for which $p(Y) = \sum_{i \in Y} p_i = 0.5$, then, due to the definition (3.5.17), for the original values s_i, we will have

$$\sum_{i \in Y} s_i = 0.5 \cdot \sum_{j=1}^{n} s_j. \qquad (3.5.18)$$

This is exactly the solution to the subset sum problem. Vice versa, if we have a set Y for which the equality (3.5.18) is satisfied, then for the probabilities (3.5.17) we get $p(Y) = 0.5$.

Conclusion. The reduction shows that in the probabilistic case, the problem of coming up with a good question is indeed NP-hard.

Fuzzy approach to describing uncertainty: a description. In the fuzzy approach, we assign, to each variant i, its degree of possibility. The resulting fuzzy values are usually *normalized*, so that the largest of these values if equal to 1: $\max_i \mu_i = 1$; see, e.g., [65, 120, 163].

How to quantify amount of uncertainty: fuzzy case. In the case of fuzzy uncertainty, one of the most widely used ways to gauge uncertainty is to use an expression

$$S = \sum_{i=1}^{n} f(\mu_i), \qquad (3.5.19)$$

for some strictly increasing continuous function $f(z)$ for which $f(0) = 0$; see, e.g., [123].

This is the amount that we want to decrease by asking an appropriate question.

How do we select a question: idea. We would thus like to find the question that maximally decreases the uncertainty. Since in the fuzzy case, uncertainty is measured by the expression (3.5.19), we thus want to find a question that maximally decreases the value of this expression.

How the answer changes the entropy. Once we know the answer to our "yes"-"no" question, the degrees of belief μ_i change.

If the answer was "yes", this means that the variants from the "no"-set N are no longer possible. For such variants $i \in N$, the new degrees are 0s: $\mu_i' = 0$. For variants from the "yes"-set Y, the new degree can be obtained by one of two different ways (see, e.g., [5–7, 16, 17, 27–29, 50]):

- in the *numerical approach*, we normalize the remaining degrees so that the maximum is equal to 1, i.e., we take

$$\mu_i' = \frac{\mu_i}{\max\limits_{j \in Y} \mu_j};$$

(3.5.20)

- in the *ordinal approach*, we raise the largest values to 1, while keeping the other values unchanged:

$$\mu_i' = 1 \text{ if } \mu_i = \max_{j \in Y} \mu_j;$$

(3.5.21)

$$\mu_i' = \mu_i \text{ if } \mu_i < \max_{j \in Y} \mu_j.$$

(3.5.22)

Based on the new values μ_i', we compute the new complexity value

$$S' = \sum_{i \in Y} f(\mu_i').$$

(3.5.23)

On the other hand, if the answer was "no", this means that the variants from the "yes"-set Y are no longer possible. For such variants $i \in Y$, the new degrees are 0s: $\mu_i'' = 0$. For variants from the "no"-set N, the new degree can be obtained by one of the same two different ways as in the case of the "yes" answer:

- in the *numerical approach*, we normalize the remaining degree so that the maximum is equal to 1, i.e., we take

$$\mu_i'' = \frac{\mu_i}{\max\limits_{j \in N} \mu_j};$$

(3.5.24)

- in the *ordinal approach*, we raise the largest values to 1, while keeping the other values unchanged:

$$\mu_i'' = 1 \text{ if } \mu_i = \max_{j \in N} \mu_j;$$

(3.5.25)

$$\mu_i'' = \mu_i \text{ if } \mu_i < \max_{j \in N} \mu_j.$$

(3.5.26)

Based on the new values μ_i'', we compute the new complexity value

$$S'' = \sum_{i \in N} f(\mu_i''). \tag{3.5.27}$$

In the case of the "yes" answer, the uncertainty decreases by the amount $S - S'$. In the case of the "no"-answer, the uncertainty decreases by the amount $S - S''$. In this case, we do not know the probabilities of "yes" and "no" answers, so we cannot estimate the expected decrease. What we can estimate is the *guaranteed* decrease

$$\overline{S}(Y) = \min(S - S', S - S''). \tag{3.5.28}$$

This value describes how much of a decrease we can guarantee if we use the "yes"-"no" answer corresponding to the set Y.

Thus, we arrive at the following formulation of the problem in precise terms.

Formulation of the problem in precise terms.

- We are *given* the degrees $\mu_1, \ldots, \mu_n$ for which

$$\max_i \mu_i = 1.$$

- We need to *find* a set $Y \subset \{1, \ldots, n\}$ for which the expected decrease in uncertainty $\overline{S}(Y)$ is the largest possible.

Here, $\overline{S}(Y)$ is described by the formula (3.5.28), and the components of this formula are described in formulas (3.5.19)–(3.5.27).

Comment. Strictly speaking, we need to solve two optimization problems:

- the problem corresponding to the numerical approach, and
- the problem corresponding to the ordinal approach.

Formulation of the main fuzzy-related result. Our main result is that for both approaches (numerical and ordinal) the problem of coming up with the best possible question is NP-hard.

How we prove this result. Similarly to the probabilistic case, we prove this result by reducing the subset sum problem to this problem.

Reduction. Let $s_1, \ldots, s_m$ be positive integers. To solve the corresponding subset sum problem, let us select a small number $\varepsilon > 0$ and consider the following $n = m+2$ degrees: $\mu_i = f^{-1}(\varepsilon \cdot s_i)$ for $i \leq m$ and $\mu_{m+1} = \mu_{m+2} = 1$, where $f^{-1}(z)$ denotes an inverse function to $f(z)$: $f^{-1}(z)$ is the value t for which $f(t) = z$.

For these values, we have three possible relations between the set Y and the variants $m + 1$ and $m + 2$:

- the first case is when the set Y contains both these variants;
- the second case is when the set Y contains none of these two variants, and

- the third case is when the set Y contains exactly one of these two variants.

Let us show that when ε is sufficiently small, then the largest guaranteed decrease is attained in the third case.

Indeed, one can easily check that

$$\overline{S}(Y) = \min(S - S', S - S'') = S - \max(S', S'');$$

Thus, the guaranteed decrease $\overline{S}(Y)$ is the largest when the maximum

$$\max(S', S'') \tag{3.5.29}$$

is the smallest.

In the first case, the values μ_i' contain two 1s, hence $S' = \sum_{i \in Y} f(\mu_i') \geq 2f(1)$.

Thus, $\max(S', S'') \geq 2f(1)$.

In the second case, the values μ_i'' contain two 1s, hence $S'' = \sum_{i \in N} f(\mu_i'') \geq 2f(1)$.

Thus, $\max(S', S'') \geq 2f(1)$.

In the third case, when one of the 1s is in Y and another one is in N, both sets Y and N contain 1s, so there is no need for normalization. Therefore, we have:

- $\mu_i' = \mu_i$ for $i \in Y$ and
- $\mu_i'' = \mu_i$ for $i \in N$.

Thus,

$$S' = \sum_{i \in Y} f(\mu_i) = f(1) + \sum_{i \in Y, i \leq m} f(\mu_i), \tag{3.5.30}$$

and

$$S'' = \sum_{i \in N} f(\mu_i) = f(1) + \sum_{i \in Y, i \leq m} f(\mu_i). \tag{3.5.31}$$

Due to our selection of μ_i, we have $f(\mu_i) = \varepsilon \cdot s_i$, so:

$$S' = \sum_{i \in Y} f(\mu_i) = f(1) + \sum_{i \in Y, i \leq m} \varepsilon \cdot s_i = f(1) + \varepsilon \cdot \sum_{i \in Y, i \leq m} s_i, \tag{3.5.32}$$

and

$$S'' = \sum_{i \in N} f(\mu_i) = f(1) + \sum_{i \in N, i \leq m} \varepsilon \cdot s_i = f(1) + \varepsilon \cdot \sum_{i \in N, i \leq m} s_i. \tag{3.5.33}$$

When $\varepsilon \cdot \sum_{i=1}^{m} s_i < f(1)$, we have $S' < 2f(1)$, $S'' < 2f(1)$, and therefore, $\max(S', S'') < 2f(1)$. Hence, for sufficiently small ε, the smallest possible value of the maximum (3.5.29) is indeed attained in the third case.

In this third case, due to (3.5.32) and (3.5.33), we have

$$\max(S', S'') = f(1) + \varepsilon \cdot \max\left(\sum_{i \in Y, i \le m} s_i, \sum_{i \in N, i \le m} s_i\right).$$

The sets Y and N are complementary to each other, hence

$$\sum_{i \in Y, i \le m} s_i + \sum_{i \in N, i \le m} s_i = \sum_{i=1}^{m} s_i.$$

If the two sums $\sum_{i \in Y, i \le m} s_i$ and $\sum_{i \in N, i \le m} s_i$ are different, then one of them is larger that one half of the total sum $\sum_{i=1}^{m} s_i$; thus, the maximum $\max(S', S'')$ is also larger than one half of the total sum. The only way to get the smallest possible value—exactly one half of the total sum—is when the sums are equal to each other, i.e., when each sum is exactly one half of the total sum:

$$\sum_{i \in Y, i \le m} s_i = \frac{1}{2} \cdot \left(\sum_{i=1}^{m} s_i\right).$$

This is exactly the solution to the subset problem. Thus, we have found the reduction of the known NP-hard subset sum problem to our problem of coming up with a good question—which implies that our problem is also NP-hard.

Summary. The reduction shows that in both fuzzy approaches, the problem of coming up with a good question is indeed NP-hard.

Need to consider interval-valued fuzzy sets. The usual [0, 1]-based fuzzy logic is based on the assumption that an expert can describe his or her degree of uncertainty by a number from the interval [0, 1]. In many practical situations, however, an expert is uncertain about his/her degree of uncertainty. In such situations, it is more reasonable to describe the expert's degree of certainty not by a single number, but by an interval which is a subinterval of the interval [0, 1].

Such interval-valued fuzzy techniques have indeed led to many useful applications; see, e.g., [103, 104, 115].

The problem of selecting a good question is NP-hard under interval-valued fuzzy uncertainty as well. Indeed, the usual fuzzy logic is a particular case of interval-valued fuzzy logic—when all intervals are degenerate, i.e., are of the form $[a, a]$ for a real number a. It is easy to prove that if a particular case of a problem is NP-hard, the whole problem is also NP-hard.

Thus, since the problem of selecting a good question is NP-hard for the case of usual [0, 1]-based fuzzy uncertainty, it is also NP-hard for the more general case of interval-valued fuzzy uncertainty.

3.5.2 The Result of Education: Justifications of Rasch Model

The more skills a student acquires, the more successful this student is with the corresponding tasks. Empirical data shows that the success in a task grows as a logistic function of skills; this dependence is known as the *Rasch model*. In this subsection, we provide two uncertainty-based justifications for this model: the first justification provides a simple fuzzy-based intuitive explanation for this model, while the second—more complex one—explains the exact quantitative behavior of the corresponding dependence.

The results of this subsection first appeared in [77].

Need to understand how success in a task depends on the skills level. As a student acquires more skills, this student becomes more successful in performing corresponding tasks. This is how we gauge the level of the knowledge and skills acquired by a student: by checking how well the student performs on the corresponding tasks.

For each level of student skills, the student is usually:

- very successful in solving simple problems,
- not yet successful in solving problems which are—to this student—too complex, and
- reasonably successful in solving problems which are of the right complexity.

To design adequate tests—and to adequately use the results of these tests to gauge the student's skills level—it is desirable to have a good understanding of how a success in a task depends on the student's skill level and on the problem's complexity.

How do we gauge success. In order to understand the desired dependence, we need to clarify how the success is measured. This depends on the type of the corresponding tasks.

For simple tasks, a student can either succeed in a task or not. In this case, a good measure of success is the proportion of tasks in which the student succeeded. In terms of uncertainty techniques, the resulting grade is simply the probability of success.

In more complex tasks, a student may succeed in some subtasks and fail in others. The simplest—and probably most frequent—way of gauging the student's success is to assign weights to different subtasks and to take, as a student's grade, the sum of the weights corresponding to successful subtasks. This somewhat mechanistic way of grading is fast and easy to automate, but it often lacks nuances: for example, it does not allow taking into account to what extent the student succeeded in each non-fully-successful subtask. A more adequate (and more complex) way—used, e.g., in grading essays—is to ask expert graders to take into account all the specifics of the student's answer and to come up with an appropriate overall grade. In terms of uncertainty techniques, this grade can be viewed as a particular case of a fuzzy degree [65, 120, 163].

Rasch model. Empirical data shows that, no matter how we measure the success rate, the success s in a task can be estimated by the following formula [101]:

$$s = \frac{1}{1 + \exp(c - \ell)}, \tag{3.5.34}$$

where c is an (appropriately re-scaled) complexity of the task and ℓ is an (also appropriately re-scaled) skill level of a student.

This formula was first proposed—in a general psychological context—by G. Rasch [138]; it is therefore known as the *Rasch model* [101].

The remaining challenge. While, empirically, this formula seems to work reasonably well, practitioners are somewhat reluctant to use it widely, since it lacks a deeper justification.

What we do in this subsection. In this subsection, we provide two possible justifications for the Rasch model. The first is a simple fuzzy-based justification which provides a good intuitive explanation for this model and, thus, will hopefully enhance its use in teaching practice. The second is a somewhat more sophisticated explanation which is less intuitive but provides a justification for the specific quantitative type of the dependence (3.5.34).

We are looking for a dependence: reminder. For a fixed level of the task's complexity c, we need to find out how the success rate s depends on the skill level ℓ. In other words, we need to find a function $g(\ell)$ for which $s = g(\ell)$.

Skill level. In general, the skill level ℓ can go from a very small number (practically $-\infty$) to a very large number (practically $+\infty$) corresponding to an extremely skilled person.

Monotonicity. The desired function $s = g(\ell)$ that describes the dependence of the success s on the skill level ℓ is clearly monotonic: the more skills a student has acquired, the larger the success in solving the tasks.

Extreme cases. In the absence of skills, a student cannot succeed in the corresponding tasks, so when $\ell \to -\infty$, we have $s = g(\ell) \to 0$. On the other side, when a person is very skilled, this person should have a perfect success in all the tasks, i.e., we should have $s = g(\ell) \to 1$ when $\ell \to +\infty$.

Smoothness. Intuitively, a very small increase in the skill level ℓ can also result in a very small increase in the success s. Thus, it is reasonable to assume that the desired dependence $s = g(\ell)$ is differentiable (smooth).

Let us make the problem easier. Let us use smoothness to reformulate the problem of determining the dependence $s = g(\ell)$ so that it will be easier to process in a computer and easier to describe in directly measurable terms.

Towards making the problem easier to process in a computer. If we change the skills ℓ a little bit, to $\ell + \Delta$, the success rate changes also a little bit. Thus, once we know the original value $s = g(\ell)$ of the success rate, and we are interested in the new value $s' = g(\ell + \Delta)$ of the success rate, it is convenient:

- not to describe the value by itself,
- but rather to describe the resulting small change $s' - s = g(\ell + \Delta\ell) - g(\ell)$ in the success rate.

This difference is smaller that the original value $g(\ell + \Delta\ell)$ and thus, requires fewer bits to record.

For small $\Delta\ell$, this difference is approximately equal to

$$\frac{dg}{d\ell} \cdot \Delta\ell.$$

Thus, describing such differences is equivalent to describing the corresponding derivative

$$\frac{dg}{d\ell}.$$

How to make the problem easier to describe in directly measurable terms. In principle, we can describe this derivative in terms of the skills level ℓ, but since the directly observable characteristic is the success s, it is more convenient to express the derivative

$$\frac{dg}{d\ell} = f(s) = f(g(\ell))$$

for an appropriate function $f(s)$.

Let us now use our understanding of this problem to describe this function $f(s)$.

Let us describe our intuition about $f(s)$ in imprecise ("fuzzy") terms. When there are no skills, i.e., when $\ell \approx -\infty$ and $s = g(\ell) \approx 0$, adding a little bit of skills does not help much. So, when s is small, the derivative

$$\frac{dg}{d\ell}$$

is also small. In other words, the derivative

$$\frac{dg}{d\ell}$$

is reasonable if and only if $s = g(\ell)$ is not small (i.e., reasonably big).

On the other hand, when s is really big, i.e., $s = g(\ell) \approx 1$, then the student is already able to solve the corresponding tasks, and adding skills does not change much in this ability. So, for the derivative

$$\frac{dg}{d\ell}$$

to be reasonable, the value $s = g(\ell)$ must be big, but not too big.

From a fuzzy description to a reasonable crisp equation. The derivative is reasonable when s is big but not too big. In this context, "but" means "and", so the degree to which this rule is applicable can be estimated as the degree to which s is big *and* s is not too big.

How can we describe "big" in this context? The value s is from the interval $[0, 1]$. The value 0 is clearly not big, the value 1 is clearly big. Thus, the corresponding membership function should be 0 when $s = 0$ and 1 when $s = 1$. The simplest such membership function is $\mu(s) = s$. A natural description of "not big" is thus $1 - s$. If we use product for "and"—one of the most widely used "and"-operations in fuzzy logic—we conclude that the degree to which the derivative

$$\frac{dg}{d\ell}$$

is reasonable is $s \cdot (1 - s) = g(\ell) \cdot (1 - g(\ell))$. Thus, we arrive at the equation

$$\frac{dg}{d\ell} = g \cdot (1 - g). \tag{3.5.35}$$

Solving this equation leads exactly to the Rasch model. To solve the equation (3.5.35), let us move all the terms containing the unknown function s to one side and all other terms to another side. Thus, we get

$$\frac{dg}{g \cdot (1 - g)} = d\ell. \tag{3.5.36}$$

One can easily check that the fraction

$$\frac{1}{g \cdot (1 - g)}$$

can be represented as the sum

$$\frac{1}{g \cdot (1 - g)} = \frac{1}{g} + \frac{1}{1 - g}.$$

Thus, the Eq. (3.5.36) has the form

$$\frac{dg}{g} + \frac{dg}{1 - g} = d\ell. \tag{3.5.37}$$

Integrating both sides, we conclude that

$$\ln(g) - \ln(1 - g) = \ell - c, \tag{3.5.38}$$

for some constant c. Thus, $\ln(1 - g) - \ln(g) = c - \ell$. Exponentiating both sides, we get

$$\frac{1 - g}{g} = \exp(c - \ell),$$

i.e.,

$$\frac{1}{g} - 1 = \exp(c - \ell).$$

Thus,

$$\frac{1}{g} = 1 + \exp(c - \ell)$$

and

$$g(\ell) = \frac{1}{1 + \exp(c - \ell)}$$

for some parameter c. This is exactly the Rasch model.

Comment. What if we use a different "and"-operation, for example, $\min(a, b)$? Let us show that in this case, we also get a meaningful model.

Indeed, in this case, the corresponding equation takes the form

$$\frac{dg}{d\ell} = \min(g, 1 - g).$$

For $s = g(\ell) \le 0.5$, this leads to

$$\frac{dg}{d\ell} = g,$$

i.e., to $g(\ell) = C_- \cdot \exp(\ell)$ for some constant C_-. For $s = g(\ell) \ge 0.5$, this formula results in

$$\frac{dg}{d\ell} = 1 - g,$$

i.e., in $g(\ell) = 1 - C_+ \cdot \exp(-\ell)$ for some constant C_+. In particular, for $C_- = 0.5$, the resulting function is a cumulative distribution corresponding to the Laplace distribution, with the probability density

$$\rho(x) = \frac{1}{2} \cdot \exp(-|x|).$$

This distribution is used in many application areas—e.g., to modify the data in large databases to promote privacy; see, e.g., [30].

We need a quantitative justification. In the previous text, we provided a justification for the Rasch model, but this justification was more on the *qualitative* side. For example, to get the exact formula of the Rasch model, we used the product "and"-

operation, and we mentioned that if we use a different "and"-operation—for example, minimum—then we get a different formula (still reasonable but different).

It is therefore still necessary to provide a *quantitative* justification for the Rasch model. Let us provide this justification. It will be less simple and less intuitive that the previous qualitative justification, but it will enable us to come up with a quantitative explanation for the Rasch model.

Assumption. Let us assume that such success s depends on how much the skills level ℓ exceeds the complexity c of the task, i.e., that that success s depends on the difference $\ell - c$: $s = h(\ell - c)$ for some function $h(x)$.

Success as a measure of skills level. As we have mentioned, success in solving problems of given time is a directly observable measure of the student's skills. Thus, we can use the value $h(\ell - c)$ for some fixed c to gauge these skills.

As a result, we get different scales. Depending on which task complexity c we select, we get different numerical values describing the same skills level: if $c \neq c'$, then we get $h(\ell - c) \neq h(\ell - c')$. In other words, we have different scales for measuring the same quantity.

This is similar to scales in physics. The fact that we have different scales for measuring the same quantity is not surprising: in physics, we also have different scales depending on which starting point we use for measurement and what measuring unit we use. For example, we can measure length in inches or in centimeters, we can measure temperature in the Celsius (C) scale or in the Fahrenheit (F) scale, etc. These are all examples of different scales for measuring the same physical quantity.

Re-scaling in physics. In physics, if we change a measuring unit to a one which is a times smaller, then the corresponding numerical value multiplies by a. In other words, instead of the original numerical value x, we get a new numerical value $x' = a \cdot x$. For example, if we replace meters with centimeters, then all numerical values get multiplied by 100: 2 m becomes 200 cm.

Similarly, when we change a starting point to one which is b units smaller, the numerical value is changed by the addition to b: $x' = x + b$.

In general, if we change both the measuring unit and the starting point, we get a linear transformation $x' = a \cdot x + b$.

Physical re-scalings form a finite-dimensional transformation group. If we first apply one linear transformation, and after that another one, we still get a linear transformation. In mathematical terms, this means that the class of linear transformations is closed under composition.

For example, we can first change meters to centimeters, and then replace centimeters with inches. Then, the resulting transformation from meters to inches is still a linear transformation.

Also, if we have a transformation, e.g., from C to F, then the "inverse" transformation from F to C is also a linear transformation. In precise terms, this means that the class of all linear transformation is invariant under taking the inversion.

In general, a class of transformations which is closed under composition and under taking the inverse is called a *transformation group*. Thus, we can say that the class of all linear transformations is a transformation group.

To describe a linear transformation, it is sufficient to provide two real-valued parameters, a, and b. In general, transformation groups whose elements can be uniquely determined by a finite set of parameters are called *finite-dimensional*. Thus, the class of all linear transformations is a finite-dimensional transformation group.

In our case, we need non-linear transformations. In our case, we need to describe a transformation $f(s)$ that transforms the original success rate $s = h(\ell - c)$ into the new value $s' = g(\ell - c')$: $s' = f(s)$, i.e.,

$$h(\ell - c') = f(h(\ell - c)).$$

When $\ell \to -\infty$, we have

$$s = h(\ell - c') \to 0$$

and

$$s = h(\ell - c) \to 0.$$

Thus, for our function $f(s)$, we must have $f(0) = 0$.

Similarly, when $\ell \to +\infty$, we have $s = h(\ell - c') \to 1$ and $s = h(\ell - c) \to 1$. Thus, for our function $f(s)$, we must have $f(1) = 1$.

This immediately implies that the function $f(s)$ must be non-linear: the only linear function $f(s)$ for which $f(0) = 0$ and $f(1) = 1$ is the identity function $f(s) = s$. Thus, for our purpose, we need to consider non-linear re-scalings $f(s)$.

How can we describe non-linear transformations: general case. Which non-linear transformations are reasonable?

Similarly to physics, it is reasonable to require that if $F(s)$ is a reasonable re-scaling from scale A to scale B, and $G(s)$ is a reasonable re-scaling from scale B to scale C, then the transformation $G(F(s))$ from scale A directly to scale C should also be reasonable. In other words, the class of reasonable transformations must be closed under composition.

Also, if $F(s)$ is a reasonable transformation from scale A to scale B, then the inverse function $F^{-1}(s)$ is a reasonable transformation from scale B to scale A. Thus, the class of reasonable transformations should be closed under inversion.

Therefore, the class of reasonable transformations should form a transformation group.

Our goal is computations. Thus, we want to be able to describe such transformation in a computer. In a computer, at any given moment of time, we can only store finitely many real-valued parameters. Thus, it is reasonable to require that the class of all reasonable transformations is a finite-dimensional transformation group.

In general, linear transformations are also reasonable. Thus, to describe all reasonable transformations, we need to describe all finite-dimensional transformation groups that contain all linear transformations. Under certain smoothness conditions (and we have argued that in our case, the dependencies are smooth) such groups have been fully described: namely, it has been proven that all the transformations from such groups are fractionally linear, i.e., have the form

$$f(s) = \frac{a \cdot s + b}{c \cdot s + d} \tag{3.5.39}$$

for appropriate values a, b, c, and d; see, e.g., [48, 115, 149].

How can we describe non-linear transformations: our case. In our case, we have two restrictions on a re-scaling transformation $f(s)$: that $f(0) = 0$ and that $f(1) = 1$. Substituting the expression (3.5.39) into the equality $f(0) = 0$, we conclude that $b = 0$, thus

$$f(s) = \frac{a \cdot s}{c \cdot s + d}. \tag{3.5.40}$$

We cannot have $d = 0$, since then we would have

$$f(s) = \frac{a}{c} = \text{const}$$

for all s. Thus, $d \neq 0$.

Since $d \neq 0$, we can divide both numerator and denominator by d, and get a formula

$$f(s) = \frac{A \cdot s}{1 + C \cdot s}, \tag{3.5.41}$$

where we denoted

$$A = \frac{a}{d}$$

and

$$C = \frac{c}{d}.$$

Substituting the expression (3.5.41) into the equality $f(1) = 1$, we conclude that $A = 1 + C$, so we conclude that in our case, non-linear transformations have the form

$$f(s) = \frac{(1 + C) \cdot s}{1 + C \cdot s}. \tag{3.5.42}$$

Resulting equation for the desired dependence $s = h(\ell - c)$. The function $f(s)$ is a transformation that transforms, for two different values $c \neq c'$, the estimate $s = h(\ell - c)$ into the estimate $s' = h(\ell - c')$: for every ℓ, we have $h(\ell - c') = f(h(\ell - c))$. In particular, for $\ell = x$, $c = 0$ and $c' = -c_0$, we have $h(x + c_0) = f(h(x))$. Substituting the expression (3.5.42) for the transformation $f(s)$ into this formula, we get the following equation:

$$h(x + c_0) = \frac{(1 + C(c_0)) \cdot h(x)}{1 + C(c_0) \cdot h(x)}, \tag{3.5.43}$$

for some C which, in general depends on c_0. To find the desired dependence $h(x)$, we thus need to solve this equation.

Solving the resulting equation. Let us first simplify the Eq. (3.5.43), by taking the reciprocal (1 over) of both sides:

$$\frac{1}{h(x + c_0)} = \frac{1 + C(c_0) \cdot h(x)}{(1 + C(c_0)) \cdot h(x)} = \frac{1}{1 + C(c_0)} \cdot \frac{1}{h(x)} + \frac{C(c_0)}{1 + C(c_0)}.$$

Subtracting 1 from both sides, we get

$$\frac{1}{h(x + c_0)} - 1 = \frac{1}{1 + C(c_0)} \cdot \frac{1}{h(x)} + \frac{C(c_0)}{1 + C(c_0)} - 1 =$$

$$\frac{1}{1 + C(c_0)} \cdot \frac{1}{h(x)} - \frac{1}{1 + C(c_0)} = \frac{1}{1 + C(c_0)} \cdot \left(\frac{1}{h(x)} - 1 \right).$$

Thus, for

$$S(x) \stackrel{\text{def}}{=} \frac{1}{h(x)} - 1$$

and

$$A(c_0) \stackrel{\text{def}}{=} \frac{1}{1 + C(c_0)},$$

we get a simplified equation

$$S(x + c_0) = A(c_0) \cdot S(x). \tag{3.5.44}$$

This equation holds for all real values x and c_0.

Since the function $s = h(x)$ is differentiable, the function $S(x)$ is also differentiable and therefore, the ratio

$$A(c_0) = \frac{S(x + c_0)}{S(x)}$$

is differentiable. Differentiating both sides of the Eq. (3.5.44) with respect to c_0 and taking $c_0 = 0$, we get

$$\frac{dS}{dx} = k \cdot S,$$

where we denoted

$$k \stackrel{\text{def}}{=} \frac{dA}{dx}_{|x=0}.$$

By moving all the terms related to S to one side and all other terms to another side, we get

$$\frac{dS}{S} = k \cdot dx.$$

Integrating, we then get $\ln(S(x)) = k \cdot x + c_1$ for some integration constant c. Exponentiating both sides, we get $S(x) = \exp(k \cdot x + c_1)$. For

$$c \stackrel{\text{def}}{=} -\frac{c_1}{k},$$

we have

$$S(x) = \exp(k \cdot (x - c)). \tag{3.5.45}$$

From

$$S(x) = \frac{1}{h(x)} - 1,$$

we conclude that

$$\frac{1}{h(x)} = S(x) + 1$$

and

$$h(x) = \frac{1}{1 + S(x)},$$

i.e., in view of the formula (3.5.45):

$$h(\ell) = \frac{1}{1 + \exp(k \cdot (\ell - c))}. \tag{3.5.46}$$

A final linear re-scaling leads to the desired formula. The formula (3.5.46) is almost the formula we need, with the only difference that now we have an additional parameter k. From the requirement that the function $s = h(\ell)$ be increasing, we conclude that $k < 0$, so

$$h(\ell) = \frac{1}{1 + \exp(|k| \cdot (c - \ell))}. \tag{3.5.47}$$

We can transform the formula (3.5.47) into exactly the desired formula if we change the measuring units for both ℓ and c to a unit which is $|k|$ times smaller. In the new units, $\ell' = |k| \cdot \ell$ and $c' = |k| \cdot c$, so the formula (3.5.47) takes the desired form

$$s = h(\ell') = \frac{1}{1 + \exp(c' - \ell')}.$$

Thus, the Rasch model has indeed been justified.

Summary. It has been empirically shown that, once we know the complexity c of a task, and the skill level ℓ of a student attempting this task, the student's success s is determined by the formula

$$s = \frac{1}{1 + \exp(c - \ell)}.$$

This formula is known as the *Rasch model* since it was originally proposed—in a general psychological context—by G. Rasch. In this subsection, we provided two uncertainty-based justifications for this model:

- a simpler fuzzy-based justification that provides an intuitive semi-qualitative explanation for this formula, and
- a more complex justification that provides a quantitative explanation for the Rasch model.

Chapter 4
Towards Explaining Heuristic Techniques (Such as Fuzzy) in Expert Decision Making

In the previous chapters, we showed that bounded rationality can explain different aspects of human decision making, including seemingly irrational ones. In this chapter, we show that similar arguments can explain the success of heuristic techniques in expert decision making.

4.1 Discrete Heuristics: Case of Concept Analysis

In many practical situations, it is necessary to describe an image in words. From the purely logical viewpoint, to describe the same object, we can use concepts of different levels of abstraction: e.g., when the image is of a German Shepherd dog, we can say that it is a dog, or that it is a mammal, or that it is a German Shepherd. In such situations, humans usually select a concept which, to them, in the most natural; this concept is called the *basic level* concept. However, the notion of a basic level concept is difficult to describe in precise terms; as a result, computer systems for image analysis are not very good in selecting concepts of basic level. At first glance, since the question is how to describe human decisions, we should use notions from a (well-developed) decision theory—such as the notion of utility. However, in practice, a well-founded utility-based approach to selecting basic level concepts is not as efficient as a purely heuristic "similarity" approach. In this section, we explain this seeming contradiction by showing that the similarity approach can be actually explained in utility terms—if we use a more accurate description of the utility of different alternatives.

The results from this section first appeared in [95].

What are basic level concepts and why their are important. With the development of new algorithms and faster hardware, computer systems are getting better and better

© Springer International Publishing AG 2018
J. Lorkowski and V. Kreinovich, *Bounded Rationality in Decision Making Under Uncertainty: Towards Optimal Granularity*, Studies in Systems, Decision and Control 99, DOI 10.1007/978-3-319-62214-9_4

in analyzing images. Computer-based systems are not yet perfect, but in many cases, they can locate human beings in photos, select photos in which a certain person of interest appears, and perform many other practically important tasks.

In general, computer systems are getting better and better in performing well-defined image understanding tasks. However, such systems are much less efficient in more open-ended tasks, e.g., when they need to describe what exactly is described by a photo.

For example, when we present, to a person, a photo of a dog and ask: "What is it?", most people will say "It is a dog". This answer comes natural to us, but, somewhat surprisingly, it is very difficult to teach this answer to a computer. The problem is that from the purely logical viewpoint, the same photo can be characterized on a more abstract level ("an animal", "a mammal") or on a more concrete level ("German shepherd"). In most situations, out of many possible concepts characterizing a given object, concepts of different levels of generality, humans select a concept of a certain intermediate level. Such preferred concepts are known as *basic level* concepts.

We need to describe basic level concepts in precise terms. Detecting basic level concepts is very difficult for computers. The main reason for this difficulty is that computers are algorithmic machines. So, to teach computers to recognize basic level concepts, we need to provide an explanation of this notion in precise terms—and we are still gaining this understanding.

Current attempts to describe basic level concepts in precise terms: a brief description. When we see a picture, we make a decision which of the concepts to select to describe this picture. In decision making theory, it is known that a consistent decision making can be described by *utility theory*, in which to each alternative A, we put into correspondence a number $u(A)$ called its *utility* in such a way that a utility of a situation in which we have alternatives A_i with probabilities p_i is equal to $\sum p_i \cdot u(A_i)$; see a detailed description in Chap. 1.

Naturally, researchers tried to use utility theory to explain the notion of basic level concepts; see, e.g., [24, 45, 169]. In this approach, researchers analyze the effect of different selections on the person's behavior, and come up with the utility values that describes the resulting effects. The utility-based approach describes the basic level concepts reasonably well, but not perfectly. Somewhat surprisingly, a different approach—called *similarity approach*—seem to be more adequate in describing basic level concepts. The idea behind this approach was proposed in informal terms in [140] and has been described more formally in [109]. Its main idea is that in a hierarchy of concepts characterizing a given object, a basic level concept is the one for which the degree of similarity between elements is much higher than for the more abstract (more general) concepts and slightly smaller than for the more concrete (more specific) concepts. For example, we select a dog as a basic level concept because the degree of similarity between different dogs is much larger than similarity between different mammals—but, on the other hand, the degree of similarity between different German Shepherds is not that much higher than the degree of similarity between dogs of various breeds.

The papers [14, 15] transformed somewhat informal psychological ideas into a precise algorithms and showed that the resulting algorithms are indeed good in detecting basic level concepts.

Challenging question. From the pragmatic viewpoint, that we have an approach that works well is good news. However, from the methodological viewpoint, the fact that a heuristic approach works better than a well-founded approach based on decision theory—which describes rational human behavior—is a challenge.

What we do in this section: main result. In this section, we show—on the qualitative level—that the problem disappears if we describe utility more accurately: under this more detailed description of utility, the decision-making approach leads to the above-mentioned similarity approach.

What we do in this section: auxiliary result. It is usually more or less clear how to define degree of similarity—or, equivalent, degree of dissimilarity ("distance" $d(x, y)$) between two objects. There are several possible approaches to translate this distance between *objects* into distance between *concepts* (classes of objects). We can use worst-case distance $d(A, B)$ defined as the maximum of all the values $d(x, y)$ for all $x \in A$ and $y \in B$. Alternatively, we can use average distance as the arithmetic average of all the corresponding values $d(x, y)$. In [14], we compared these alternatives; it turns out that the average distance leads to the most adequate description of the basic level concepts.

In this section, we provide a (qualitative) explanation of this empirical fact as well.

What is the utility associated with concepts of different levels of generality. In the ideal world, when we make a decision in a certain situation, we should take into account all the information about this situation, and we should select the best decision based on this situation.

In practice, our ability to process information is limited. As a result, instead of taking into account all possible information about the object, we use a word (concept) to describe this notion, and then we make a decision based only on this word: e.g., a tiger or a dog. Instead of taking into account all the details of the fur and of the face, we decide to run away (if it is a tiger) or to wave in a friendly manner (if it is a dog).

In other words, instead of making an optimal decision for each object, we use the same decision based on an "average" object from the corresponding class. Since we make a decision without using all the information, based only on an approximate information, we thus lose some utility; see, e.g., Sect. 3.2 for a precise description of this loss.

From this viewpoint, the smaller the classes, the less utility we lose. This is what was used in the previous utility-based approaches to selecting basic level concepts.

However, if the classes are too small, we need to store and process too much information—and the need to waste resources (e.g., time) to process all this additional information also decreases utility. For example, instead of coming up with strategies corresponding to a few basic animals, we can develop separate strategies for short tigers, medium size tigers, larger tigers, etc.—but this would take more processing time and use memory resources which may be more useful for other tasks. While this is a concern, we should remember that we have billions of neurons, enough to store and process huge amounts of information, so this concern is rather secondary in comparison with a difference between being eaten alive (if it is a tiger) or not (if it is a dog).

How to transform the above informal description of utility into precise formulas and how this leads to the desired explanations. The main reason for *disutility* (loss of utility) is that in a situation when we actually have an x, we use an approach which is optimal for a similar (but slightly different) object y. For example, instead of making a decision based on observing a very specific dog x, we ignore all the specifics of this dog, and we make a decision based only one the fact that x is a dog, i.e., in effect, we make a decision based on a "typical" dog y.

The larger the distance $d(x, y)$ between the objects x and y, the larger this disutility U. Intuitively, different objects within the corresponding class are similar to each other— otherwise they would not be classified into the same class. Thus, the distance $d(x, y)$ between objects from the same class are small. We can therefore expand the dependence of U on $d(x, y)$ in Taylor series and keep only the first few terms in this dependence. In general, $U = a_0 + a_1 \cdot d + a_2 \cdot d^2 + \ldots$ When the distance is 0, i.e., when $x = y$, there is no disutility, so $U = 0$. Thus, $a_0 = 0$ and the first non-zero term in the Taylor expansion is $U \approx a_1 \cdot d(x, y)$.

Once we act based on the class label ("concept"), we only know that an object belongs to the class, we do not know the exact object within the class. We may have different objects from this class with different probabilities. By the above property of utility, the resulting disutility of selecting a class is equal to the *average* value of the disutility—and is, thus proportional to the *average distance* $d(x, y)$ between objects from a given class. *This explains why average distance works better then the worst-case distance.*

When we go from a more abstract concept (i.e., from a larger class) to a more specific concept (i.e., to a smaller class of objects), the average distance decreases— and thus, the main part U_m of disutility decreases: $U'_m < U_m$. However, as we have mentioned, in addition to this main part of disutility U_m, there is also an additional secondary (smaller) part of utility $U_s \ll U_m$, which increases when we go to a more specific concept: $U'_s > U_s$.

On the qualitative level, this means the following: if the less general level has a much smaller degree of similarity (i.e., a drastically smaller average distance between the objects on this level), then selecting a concept on this less general level drastically decreases the disutility $U'_m \ll U_m$, and this decrease $U_m - U'_m \gg 0$ overwhelms the (inevitable) increase $U'_s - U_s$ in the secondary part of disutility, so that $U' = U_m + U'_s < U_m + U_s = U$. On the other hand, if the decrease in degree of similarity is small (i.e., $U'_m \approx U_m$), the increase in the secondary part of disutility $U'_s - U_s$ can over-stage the small decrease $U'_m - U_m$.

A basic level concept is a concept for which disutility U' is smaller than for a more general concept U and smaller than for a more specific concept U''. In view of the above, this means that there should be a drastic difference between the degree of similarity U'_m at this level and the degree of similarity U_m at the more general level—otherwise, on the current level, we would not have smaller disutility. Similarly, there should be a small difference between the degree of similarity at the current level U'_m and the degree of similarity U''_m at the more specific level—otherwise, on the current level, we would not have smaller disutility. *This explains the similarity approach in utility terms.*

4.2 Continuous Heuristics: Case of Fuzzy Techniques

One of the main methods for eliciting the values of the membership function $\mu(x)$ is to use the Likert-type scales, i.e., to ask the user to mark his or her degree of certainty by an appropriate mark k on a scale from 0 to n and take $\mu(x) = k/n$. In this section, we show how to describe this process in terms of the traditional decision making. Our conclusion is that the resulting membership degrees incorporate both probability and utility information. It is therefore not surprising that fuzzy techniques often work better than probabilistic techniques—which only take into account the probability of different outcomes.

The results from this section appeared in [86, 87, 90].

4.2.1 Fuzzy Uncertainty: A Brief Description

Fuzzy uncertainty: a usual description. Fuzzy logic (see, e.g., [65, 120, 163]) has been designed to describe imprecise ("fuzzy") natural language properties like "big", "small", etc. In contrast to "crisp" properties like $x \leq 10$ which are either true or false, experts are not 100% sure whether a given value x is big or small. To describe such properties P, fuzzy logic proposes to assign, to each possible value x, a degree $\mu_P(x)$ to which the value x satisfies this property:

- the degree $\mu_P(x) = 1$ means that we are absolutely sure that the value x satisfies the property P;
- the degree $\mu_P(x) = 0$ means that we are absolutely sure that the value x does not satisfy the property P; and
- intermediate degrees $0 < \mu_P(x) < 1$ mean that we have *some* confidence that x satisfies the property P but we also have a certain degree of confidence that the value x does not satisfy this property.

How do we elicit the degree $\mu_P(x)$ from the expert? One of the usual ways is to use a *Likert-type scale*, i.e., to ask the expert to mark his or her degree of confidence that the value x satisfies the property P by one of the marks $0, 1, \ldots, n$ on a scale from 0 to n. If an expert marks m on a scale from 0 to n, then we take the ratio m/n as the desired degree $\mu_P(x)$. For example, if an expert marks her confidence by a value 7 on a scale from 0 to 10, then we take $\mu_P(x) = 7/10$.

For a fixed scale from 0 to n, we only get $n + 1$ values this way: $0, 1/n, 2/n, \ldots,$ $(n - 1)/n = 1 - 1/n$, and 1. If we want a more detailed description of the expert's uncertainty, we can use a more detailed scale, with a larger value n.

Traditional decision making theory: a brief reminder. Decision making has been analyzed for decades. Efficient models have been developed and tested to describe human decision making, and the resulting tools have been effectively used in business and in other decision areas; see, e.g., [35, 36, 63, 97, 133]. These models are not perfect—this is one the reasons why fuzzy methods are needed—but these tools provide a reasonable first-order approximation description of human decision making.

Need to combine fuzzy techniques and traditional decision making techniques, and the resulting problem that we solve in this section. Traditional decision making tools are useful but have their limitations. Fuzzy tools are also known to be very useful, in particular, they are known to be useful in control and in decision making (see, e.g., [65, 120]), so a natural idea is to combine these two techniques.

To enhance this combination, it is desirable to be able to describe both techniques in the same terms. In particular, it is desirable to describe fuzzy uncertainty in terms of traditional decision making. To the best of our knowledge, this has not been done before; we hope that our description will lead to useful applications in practical decision making.

4.2.2 How to Describe Selection on a Likert-Type Scale in Terms of Traditional Decision Making

How do we place marks on a Likert-type scale? We would like to find out how people decide to mark some values with different labels on a Likert-type scale. To understand this, let us recall how this marking is done. Suppose that we have Likert-type scale with $n + 1$ labels $0, 1, 2, \ldots, n$, ranging from the smallest to the largest.

Then, if the actual value of the quantity x is very small, we mark label 0. At some point, we change to label 1; let us mark this threshold point by x_1. When we continue increasing x, we first have values marked by label 1, but eventually reach a new threshold after which values will be marked by label 2; let us denote this threshold by x_2, etc. As a result, we divide the range $[\underline{X}, \overline{X}]$ of the original variable into $n + 1$ intervals $[x_0, x_1], [x_1, x_2], \ldots, [x_{n-1}, x_n], [x_n, x_{n+1}]$, where $x_0 = \underline{X}$ and $x_{n+1} = \overline{X}$:

- values from the first interval $[x_0, x_1]$ are marked with label 0;
- values from the second interval $[x_1, x_2]$ are marked with label 1;
- …
- values from the n-th interval $[x_{n-1}, x_n]$ are marked with label n;
- values from the $(n + 1)$-st interval $[x_n, x_{n+1}]$ are marked with label $n + 1$.

Then, when we need to make a decision, we base this decision only on the label, i.e., only on the interval to which x belongs. In other words, we make n different decisions depending on whether x belongs to the interval $[x_0, x_1]$, to the interval $[x_1, x_2]$, ..., or to the interval $[x_n, x_{n+1}]$.

Decisions based on the Likert-type discretization are imperfect. Ideally, we should take into account the exact value of the variable x. When we use Likert-type scale, we only take into account an interval containing x and thus, we do not take into account part of the original information. Since we only use part of the original information about x, the resulting decision may not be as good as the decision based on the ideal complete knowledge.

For example, an ideal office air conditioner should be able to maintain the exact temperature at which a person feels comfortable. People are different, their temperature preferences are different, so an ideal air conditioner should be able to maintain any temperature value x within a certain range $[\underline{X}, \overline{X}]$. In practice, some air conditioners only have a finite number of settings. For example, if we have setting corresponding to 65, 70, 75, and 80° then a person who prefers 72° will probably select the 70 setting or the 75 setting. In both cases, this person will be somewhat less comfortable than if there was a possibility of an ideal 72° setting.

How do we select a Likert-type scale: main idea. According to the general ideas of traditional (utility-based) approach to decision making, we should select a Likert scale for which the expected utility is the largest.

To estimate the utility of decisions based on each scale, we will take into account the just-mentioned fact that decisions based on the Likert-type discretization are imperfect. In utility terms, this means that the utility of the Likert-based decisions is, in general, smaller than the utility of the ideal decision.

Which decision should we choose within each label? In the ideal situation, if we could use the exact value of the quantity x, then for each value x, we would select an optimal decision $d(x)$, a decision which maximizes the person's utility.

If we only know the label k, i.e., if we only know that the actual value x belongs to the k-th interval $[x_k, x_{k+1}]$, then we have to make a decision based only on this information. In other words, we have to select one of the possible values $\widetilde{x}_k \in [x_k, x_{k+1}]$, and then, for all x from this interval, use the decision $d(\widetilde{x}_k)$ based on this value.

Which value $\widetilde{x}_k$ should we choose: idea. According to the traditional approach to decision making, we should select a value for which the expected utility is the largest.

Which value $\widetilde{x}_k$ should we choose: towards a precise formulation of the problem. To find this expected utility, we need to know two things:

- we need to know the probability of different values of x; these probabilities can be described, e.g., by the probability density function $\rho(x)$;
- we also need to know, for each pair of values x' and x, what is the utility $u(x', x)$ of using a decision $d(x')$ in the situation in which the actual value is x.

In these terms, the expected utility of selecting a value $\widetilde{x}_k$ can be described as

$$\int_{x_k}^{x_{k+1}} \rho(x) \cdot u(\widetilde{x}_k, x) \, dx. \tag{4.2.1}$$

Thus, for each interval $[x_k, x_{k+1}]$, we need to select a decision $d(\tilde{x}_k)$ corresponding to the value $\tilde{x}_k$ for which the expression (4.2.1) attains its largest possible value. The resulting expected utility is equal to

$$\max_{\tilde{x}_k} \int_{x_k}^{x_{k+1}} \rho(x) \cdot u(\tilde{x}_k, x)\, dx. \tag{4.2.2}$$

How to select the best Likert-type scale: general formulation of the problem. The actual value x can belong to any of the $n+1$ intervals $[x_k, x_{k+1}]$. Thus, to find the overall expected utility, we need to add the values (4.2.2) corresponding to all these intervals. In other words, we need to select the values $x_1, \ldots, x_n$ for which the following expression attains its largest possible value:

$$\sum_{k=0}^{n} \max_{\tilde{x}_k} \int_{x_k}^{x_{k+1}} \rho(x) \cdot u(\tilde{x}_k, x)\, dx. \tag{4.2.3}$$

Equivalent reformulation in terms of disutility. In the ideal case, for each value x, we should use a decision $d(x)$ corresponding to this value x, and gain utility $u(x, x)$. In practice, we have to use decisions $d(x')$ corresponding to a slightly different value, and thus, get slightly worse utility values $u(x', x)$. The corresponding decrease in utility $U(x', x) \overset{\text{def}}{=} u(x, x) - u(x', x)$ is usually called *disutility*. In terms of disutility, the function $u(x', x)$ has the form

$$u(x', x) = u(x, x) - U(x', x),$$

and thus, the optimized expression (4.2.1) takes the form

$$\int_{x_k}^{x_{k+1}} \rho(x) \cdot u(x, x)\, dx - \int_{x_k}^{x_{k+1}} \rho(x) \cdot U(\tilde{x}_k, x)\, dx.$$

The first integral does not depend on $\tilde{x}_k$; thus, the expression (4.2.1) attains its maximum if and only if the second integral attains its minimum. The resulting maximum (4.2.2) thus takes the form

$$\int_{x_k}^{x_{k+1}} \rho(x) \cdot u(x, x)\, dx - \min_{\tilde{x}_k} \int_{x_k}^{x_{k+1}} \rho(x) \cdot U(\tilde{x}_k, x)\, dx. \tag{4.2.4}$$

Thus, the expression (4.2.3) takes the form

$$\sum_{k=0}^{n} \int_{x_k}^{x_{k+1}} \rho(x) \cdot u(x, x)\, dx - \sum_{k=0}^{n} \min_{\tilde{x}_k} \int_{x_k}^{x_{k+1}} \rho(x) \cdot U(\tilde{x}_k, x)\, dx.$$

The first sum does not depend on selecting the thresholds. Thus, to maximize utility, we should select the values $x_1, \ldots, x_n$ for which the second sum attains its smallest possible value:

$$\sum_{k=0}^{n} \min_{\tilde{x}_k} \int_{x_k}^{x_{k+1}} \rho(x) \cdot U(\tilde{x}_k, x) \, dx \to \min. \tag{4.2.5}$$

Let us recall that we are interested in the membership function. For a general Likert-type scale, we have a complex optimization problem (4.2.5). However, we are not interested in general Likert-type scales per se, what we are interested in is the use of Likert-type scales to elicit the values of the membership function $\mu(x)$.

As we have mentioned in Sect. 1, in an n-valued scale:

- the smallest label 0 corresponds to the value $\mu(x) = 0/n$,
- the next label 1 corresponds to the value $\mu(x) = 1/n$,
- ...
- the last label n corresponds to the value $\mu(x) = n/n = 1$.

Thus, for each n:

- values from the interval $[x_0, x_1]$ correspond to the value $\mu(x) = 0/n$;
- values from the interval $[x_1, x_2]$ correspond to the value $\mu(x) = 1/n$;
- ...
- values from the interval $[x_n, x_{n+1}]$ correspond to the value $\mu(x) = n/n = 1$.

For each n, we have finitely many possible values of $\mu(x)$. The actual real-number value of the membership function $\mu(x)$ corresponds to the limit when $n \to \infty$, i.e., in effect, to very large values of n. Thus, in our analysis, we will assume that the number n of labels is huge—and thus, that the width of each of $n + 1$ intervals $[x_k, x_{k+1}]$ is very small.

Let us take into account that each interval is narrow. Let us use the fact that each interval is narrow to simplify the expression $U(x', x)$ and thus, the optimized expression (4.2.5).

In the expression $U(x', x)$, both values x' and x belong to the same narrow interval and thus, the difference $\Delta x \stackrel{\text{def}}{=} x' - x$ is small. Thus, we can expand the expression $U(x', x) = U(x + \Delta x, x)$ into Taylor series in Δx, and keep only the first non-zero term in this expansion. In general, we have

$$U(x + \Delta, x) = U_0(x) + U_1 \cdot \Delta x + U_2(x) \cdot \Delta x^2 + \ldots,$$

where

$$U_0(x) = U(x, x), \quad U_1(x) = \frac{\partial U(x + \Delta x, x)}{\partial(\Delta x)},$$

$$U_2(x) = \frac{1}{2} \cdot \frac{\partial^2 U(x + \Delta x, x)}{\partial^2(\Delta x)}. \tag{4.2.7}$$

Here, by definition of disutility, we get $U_0(x) = U(x, x) = u(x, x) - u(x, x) = 0$. Since the utility is the largest (and thus, disutility is the smallest) when $x' = x$, i.e., when $\Delta x = 0$, the derivative $U_1(x)$ is also equal to 0—since the derivative of each (differentiable) function is equal to 0 when this function attains its minimum. Thus, the first non-trivial term corresponds to the second derivative:

$$U(x + \Delta x, x) \approx U_2(x) \cdot \Delta x^2,$$

i.e., in other words, that

$$U(\widetilde{x}_k, x) \approx U_2(x) \cdot (\widetilde{x}_k - x)^2.$$

Substituting this expression into the expression

$$\int_{x_k}^{x_{k+1}} \rho(x) \cdot U(\widetilde{x}_k, x) \, dx$$

that needs to be minimized if we want to find the optimal $\widetilde{x}_k$, we conclude that we need to minimize the integral

$$\int_{x_k}^{x_{k+1}} \rho(x) \cdot U_2(x) \cdot (\widetilde{x}_k - x)^2 \, dx. \tag{4.2.8}$$

This new integral is easy to minimize: if we differentiate this expression with respect to the unknown $\widetilde{x}_k$ and equate the derivative to 0, we conclude that

$$\int_{x_k}^{x_{k+1}} \rho(x) \cdot U_2(x) \cdot (\widetilde{x}_k - x) \, dx = 0,$$

i.e., that

$$\widetilde{x}_k \cdot \int_{x_k}^{x_{k+1}} \rho(x) \cdot U_2(x) \, dx = \int_{x_k}^{x_{k+1}} x \cdot \rho(x) \cdot U_2(x) \, dx,$$

and thus, that

$$\widetilde{x}_k = \frac{\int_{x_k}^{x_{k+1}} x \cdot \rho(x) \cdot U_2(x) \, dx}{\int_{x_k}^{x_{k+1}} \rho(x) \cdot U_2(x) \, dx}. \tag{4.2.9}$$

This expression can also be simplified if we take into account that the intervals are narrow. Specifically, if we denote the midpoint of the interval $[x_k, x_{k+1}]$ by $\overline{x}_k \stackrel{\text{def}}{=} \dfrac{x_k + x_{k+1}}{2}$, and denote $\Delta x \stackrel{\text{def}}{=} x - \overline{x}_k$, then we have $x = \overline{x}_k + \Delta x$. Expanding the corresponding expressions into Taylor series in terms of a small value Δx and keeping only main terms in this expansion, we get

$$\rho(x) = \rho(\overline{x}_k + \Delta x) = \rho(\overline{x}_k) + \rho'(\overline{x}_k) \cdot \Delta x \approx \rho(\overline{x}_k),$$

where $f'(x)$ denoted the derivative of a function $f(x)$, and

$$U_2(x) = U_2(\overline{x}_k + \Delta x) = U_2(\overline{x}_k) + U_2'(\overline{x}_k) \cdot \Delta x \approx U_2(\overline{x}_k).$$

Substituting these expressions into the formula (4.2.9), we conclude that

$$\widetilde{x}_k = \frac{\rho(\overline{x}_k) \cdot U_2(\overline{x}_k) \cdot \int_{x_k}^{x_{k+1}} x\,dx}{\rho(\overline{x}_k) \cdot U_2(\overline{x}_k) \cdot \int_{x_k}^{x_{k+1}} dx} = \frac{\int_{x_k}^{x_{k+1}} x\,dx}{\int_{x_k}^{x_{k+1}} dx} =$$

$$\frac{\frac{1}{2} \cdot (x_{k+1}^2 - x_k^2)}{x_{k+1} - x_k} = \frac{x_{k+1} + x_k}{2} = \overline{x}_k.$$

Substituting this midpoint value $\widetilde{x}_k = \overline{x}_k$ into the integral (4.2.8) and taking into account that on the k-th interval, we have $\rho(x) \approx \rho(\overline{x}_k)$ and $U_2(x) \approx U_2(\overline{x}_k)$, we conclude that the integral (4.2.8) takes the form

$$\int_{x_k}^{x_{k+1}} \rho(\overline{x}_k) \cdot U_2(\overline{x}_k) \cdot (\overline{x}_k - x)^2\,dx = \rho(\overline{x}_k) \cdot U_2(\overline{x}_k) \cdot \int_{x_k}^{x_{k+1}} (\overline{x}_k - x)^2\,dx. \quad (4.2.8a)$$

When x goes from x_k to x_{k+1}, the difference $\Delta x = x - \overline{x}_k$ between the value x and the interval's midpoint $\overline{x}_k$ ranges from $-\Delta_k$ to Δ_k, where Δ_k is the interval's half-width:

$$\Delta_k \stackrel{\text{def}}{=} \frac{x_{k+1} - x_k}{2}.$$

In terms of the new variable Δx, the integral in the right-hand side of (4.2.8a) has the form

$$\int_{x_k}^{x_{k+1}} (\overline{x}_k - x)^2\,dx = \int_{-\Delta_k}^{\Delta_k} (\Delta x)^2\,d(\Delta x) = \frac{2}{3} \cdot \Delta_k^3.$$

Thus, the integral (4.2.8) takes the form

$$\frac{2}{3} \cdot \rho(\overline{x}_k) \cdot U_2(\overline{x}_k) \cdot \Delta_k^3.$$

The problem (4.2.5) of selecting the Likert-type scale thus becomes the problem of minimizing the sum (4.2.5) of such expressions (4.2.8), i.e., of the sum

$$\frac{2}{3} \cdot \sum_{k=0}^{n} \rho(\overline{x}_k) \cdot U_2(\overline{x}_k) \cdot \Delta_k^3. \quad (4.2.10)$$

Here, $\overline{x}_{k+1} = x_{k+1} + \Delta_{k+1} = (\overline{x}_k + \Delta_k) + \Delta_{k+1} \approx \overline{x}_k + 2\Delta_k$, so $\Delta_k = (1/2) \cdot \Delta\overline{x}_k$, where $\Delta\overline{x}_k \overset{\text{def}}{=} \overline{x}_{k+1} - \overline{x}_k$. Thus, (4.2.10) takes the form

$$\frac{1}{3} \cdot \sum_{k=0}^{n} \rho(\overline{x}_k) \cdot U_2(\overline{x}_k) \cdot \Delta_k^2 \cdot \Delta\overline{x}_k. \qquad (4.2.11)$$

In terms of the membership function, we have $\mu(\overline{x}_k) = k/n$ and $\mu(\overline{x}_{k+1}) = (k+1)/n$. Since the half-width Δ_k is small, we have

$$\frac{1}{n} = \mu(\overline{x}_{k+1}) - \mu(\overline{x}_k) = \mu(\overline{x}_k + 2\Delta_k) - \mu(\overline{x}_k) \approx \mu'(\overline{x}_k) \cdot 2\Delta_k,$$

thus, $\Delta_k \approx \dfrac{1}{2n} \cdot \dfrac{1}{\mu'(\overline{x}_k)}$. Substituting this expression into (4.2.11), we get the expression $\dfrac{1}{3 \cdot (2n)^2} \cdot I$, where

$$I = \sum_{k=0}^{n} \frac{\rho(\overline{x}_k) \cdot U_2(\overline{x}_k)}{(\mu'(\overline{x}_k))^2} \cdot \Delta\overline{x}_k. \qquad (4.2.12)$$

The expression I is an integral sum, so when $n \to \infty$, this expression tends to the corresponding integral

$$I = \int \frac{\rho(x) \cdot U_2(x)}{(\mu'(x))^2} \, dx. \qquad (4.2.11a)$$

Minimizing (4.2.5) is equivalent to minimizing I. With respect to the derivative $d(x) \overset{\text{def}}{=} \mu'(x)$, we need to minimize the objective function

$$I = \int \frac{\rho(x) \cdot U_2(x)}{d^2(x)} \, dx \qquad (4.2.12a)$$

under the constraint that

$$\int_{\underline{X}}^{\overline{X}} d(x) \, dx = \mu(\overline{X}) - \mu(\underline{X}) = 1 - 0 = 1. \qquad (4.2.13)$$

By using the Lagrange multiplier method, we can reduce this constraint optimization problem to the unconstrained problem of minimizing the functional

$$I = \int \frac{\rho(x) \cdot U_2(x)}{d^2(x)} \, dx + \lambda \cdot \int d(x) \, dx, \qquad (4.2.14)$$

for an appropriate Lagrange multiplier λ. Differentiating (4.2.14) with respect to $d(x)$ and equating the derivative to 0, we conclude that $-2 \cdot \dfrac{\rho(x) \cdot U_2(x)}{d^3(x)} + \lambda = 0$, i.e., that $d(x) = c \cdot (\rho(x) \cdot U_2(x))^{1/3}$ for some constant c. Thus, $\mu(x) = \int_{\underline{X}}^{x} d(t)\, dt = c \cdot \int_{\underline{X}}^{x} (\rho(t) \cdot U_2(t))^{1/3}\, dt$. The constant c must be determined by the condition that $\mu(\overline{X}) = 1$. Thus, we arrive at the following formula (4.2.15).

Resulting formula. The membership function $\mu(x)$ obtained by using Likert-type-scale elicitation is equal to

$$\mu(x) = \frac{\int_{\underline{X}}^{x} (\rho(t) \cdot U_2(t))^{1/3}\, dt}{\int_{\underline{X}}^{\overline{X}} (\rho(t) \cdot U_2(t))^{1/3}\, dt}, \tag{4.2.15}$$

where $\rho(x)$ is the probability density describing the probabilities of different values of x, $U_2(x) \overset{\text{def}}{=} \dfrac{1}{2} \cdot \dfrac{\partial^2 U(x + \Delta x, x)}{\partial^2 (\Delta x)}$, $U(x', x) \overset{\text{def}}{=} u(x, x) - u(x', x)$, and $u(x', x)$ is the utility of using a decision $d(x')$ corresponding to the value x' in the situation in which the actual value is x.

Comment. The above formula only applies to membership functions like "large" whose values monotonically increase with x. It is easy to write a similar formula for membership functions like "small" which decrease with x. For membership functions like "approximately 0" which first increase and then decrease, we need to separately apply these formula to both increasing and decreasing parts.

Summary. The resulting membership degrees incorporate both probability and utility information. This fact *explains why fuzzy techniques often work better than probabilistic techniques*—because the probability techniques only take into account the probability of different outcomes.

Extension to interval-valued case: preliminary results and future work. In this section, we consider an ideal situation in which

- we have full information about the probabilities $\rho(x)$, and
- the user can always definitely decide between every two alternatives.

In practice, we often only have partial information about probabilities (which can be described by the intervals of possible values of $\rho(x)$) and users are often unsure which of the two alternatives is better (which can be described by interval-valued utilities).

For example, if we have no reason to believe that some values from the interval $[\underline{X}, \overline{X}]$ are more probable than others and that some values are more sensitive than others, it is natural to assume that $\rho(x) = \text{const}$ and $U_2(x) = \text{const}$, in which case the above formula (4.2.15) leads to a triangular membership function. This may *explain why triangular membership functions are successfully used in* many applications of *fuzzy* techniques.

It is desirable to extend our formulas to the general interval-valued case.

Chapter 5
Predicting How Bounded Rationality Will Affect the Quality of Future Decisions: Case Study

It is known that the problems of optimal design are NP-hard—meaning that, in general, a feasible algorithm can only produce close-to-optimal designs. The more computations we perform, the better design we can produce. In this chapter, we theoretically derive quantitative formulas describing how design quality improves with the increasing computational abilities. We then empirically confirm the resulting theoretical formula by applying it to the problem of aircraft fuel efficiency.

Results from this chapter first appeared in [80, 81].

5.1 Formulation of the Problem

Design objective is to produce an optimal design. Starting from 1980s, computers have become ubiquitous in engineering design; see, e.g., [8, 57, 70, 100]. An important breakthrough in computer-aided design was Boeing 777, the first commercial airplane which was designed exclusively by using computers; see, e.g., [141].

The main objective of a computer-aided design is to come up with a design which optimizes the corresponding objective function—e.g., fuel efficiency of an aircraft.

Optimization is, in general, NP-hard. The corresponding optimization problems are non-linear, and non-linear optimization problems are, in general, NP-hard; see, e.g., [69, 125]. This means that—under the belief of most computer scientists that $P \neq NP$—a feasible algorithm cannot always find the exact optimum; see, e.g., [42, 124]. In general, we can only find an approximate optimum.

Problem. The more computations we perform, the better the design. It is desirable to come up with a quantitative description of how increasing computational abilities improve the design quality.

© Springer International Publishing AG 2018
J. Lorkowski and V. Kreinovich, *Bounded Rationality in Decision Making Under Uncertainty: Towards Optimal Granularity*, Studies in Systems, Decision and Control 99, DOI 10.1007/978-3-319-62214-9_5

5.2   Analysis of the Problem and the Derivation of the Resulting Formula

Because of NP-hardness, more computations simply means more test cases. In principle, each design optimization problem can be solved by exhaustive search—we can try all possible combinations of parameters, and see which combination leads to the optimal design. This approach may work if we have a small number of parameters, then we can indeed try all possible combinations. If, on average, we have C possible values of each parameter, then:

- we need to compare C test cases when we have only one parameter,
- we need C^2 test cases when we have two parameters,
- and we need C^3 test cases when we have three parameters.

In general, when we have d parameters, we need to analyze C^d test cases. For large systems (e.g., for an aircraft), we have thousands of possible parameters, and for $d \approx 10^3$, the exponential value C^d exceeds the lifetime of the Universe. As a result, for realistic d, instead of the exhaustive search of *all* possible combinations of parameters, we can only test *some* combinations.

NP-hardness means, crudely speaking, that we cannot expect optimization algorithms to be significantly faster than this exponential time C^d. This means that, in effect, all possible optimization algorithms boil down to trying many possible test cases.

When computational abilities increase, we can test more cases. From this viewpoint, increasing computational abilities mean that we can test more cases. Thus, by increasing the scope of our search, we will hopefully find a better design.

How can we describe this in quantitative terms?

How to describe quality of an individual design. Since we cannot do significantly better than with a simple search, the resulting search is not well directed, we cannot meaningfully predict whether the next test case will be better or worse—because if we could, we would be able to significantly decrease the search time.

The quality of the next test case—i.e., in precise terms, the value of the objective function corresponding to the next test case—cannot be predicted and is, in this sense, a random variable.

Many different factors affect the quality of each individual design. It is known that, under reasonable conditions, the distribution of the resulting effect of several independent random factors is close to Gaussian; this fact is known as the *Central Limit Theorem*; see, e.g., [144]. Thus, we can conclude that the quality of a (randomly selected) individual design is normally distributed, with some mean μ and standard deviation σ.

What if we test n possible designs. After computation, we select the design with the largest value of the objective function. Let n denote the number of designs that our program tests. If x_i denotes the quality of the i-th design, then the resulting quality is equal to $x = \max(x_1, \ldots, x_n)$. We know that the variables x_i are independent and identically normally distributed with some mean μ and standard deviation σ. What is the resulting probability distribution for the quality x? What is the expected value of this quality?

To answer this question, let us first reduce this question to its simplest case of a standard normal distribution, with $\mu = 0$ and $\sigma = 1$. It is known that a general normally distributed random variable x_i can be represented as $x_i = \mu + \sigma \cdot y_i$. Since adding μ and multiplying by a positive constant $\sigma > 0$ does not change which of the values are larger and which are smaller, we have

$$x = \max(x_1, \ldots, x_n) = \max(\mu + \sigma \cdot y_1, \ldots, \mu + \sigma \cdot y_n) = \mu + \sigma \cdot y,$$

where $y \overset{\text{def}}{=} \max(y_1, \ldots, y_n)$.

For large n, the *max-central limit theorem* [11, 26, 31, 49] (also known as Fisher-Tippet-Gnedenko Theorem) says that the cumulative distributive function $F(y)$ for y is approximately equal to

$$F(y) \approx F_{EV}\left(\frac{y - \mu_n}{\sigma_n}\right),$$

where:

$$F_{EV}(y) \overset{\text{def}}{=} \exp(-\exp(-y))$$

is known as the *Gumbel distribution*,

$$\mu_n \overset{\text{def}}{=} \Phi^{-1}\left(1 - \frac{1}{n}\right),$$

$$\sigma_n \overset{\text{def}}{=} \Phi^{-1}\left(1 - \frac{1}{n} \cdot e^{-1}\right) - \Phi^{-1}\left(1 - \frac{1}{n}\right),$$

and $\Phi^{-1}(t)$ is the inverse function to the cumulative distribution function $\Phi(y)$ of the standard normal distribution (with mean 0 and standard deviation 1). In other words, the distribution of the random variable y is approximately equal to the distribution of the variable $\mu_n + \sigma_n \cdot \xi$, where ξ is distributed according to the Gumbel distribution.

It is known that the mean of the Gumbel distribution is equal to the Euler's constant $\gamma \approx 0.5772$. Thus, the mean value m_n of y is equal to $\mu_n + \gamma \cdot \sigma_n$. For large n, we get asymptotically

$$m_n \sim \gamma \cdot \sqrt{2 \ln(n)},$$

hence the mean value e_n of $x = \mu + \sigma \cdot y$ is asymptotically equal to

$$e_n \sim \mu + \sigma \cdot \gamma \cdot \sqrt{2 \ln(n)}.$$

Resulting formula. When we test n different cases to find the optimal design, the quality e_n of the resulting design increases with n as

$$e_n \sim \mu + \sigma \cdot \gamma \cdot \sqrt{2 \ln(n)}.$$

5.3   Case Study of Aircraft Fuel Efficiency Confirms the Theoretical Formula

Case study: brief description. As a case study, let us take the fuel efficiency of commercial aircraft; see, e.g., [71, 127, 128]. It is known that the average energy efficiency E changes with time T as

$$E = \exp(a + b \cdot \ln(T)) = C \cdot T^b,$$

for $b \approx 0.5$.

How to apply our theoretical formula to this case? The above theoretical formula $e_n \sim \mu + \sigma \cdot \gamma \cdot \sqrt{2 \ln(n)}$ describes how the quality changes with the number of computational steps n. In the case study, we know how it changes with time T. So, to compare these two formulas, we need to know how the number of computational steps which can be applied to solve the design problem changes with time T. In other words, we need to know how the computer's computational speed—i.e., the number of computational steps that a computer can perform in a fixed time period—changes with time T.

This dependence follows the known *Moore's law*, according to which the computational speed grows exponentially with time T: $n \approx \exp(c \cdot T)$ for some constant c. Crudely speaking, the computational speed doubles every two years; [53, 106].

Applying the theoretical formula to this case study. When $n \approx \exp(c \cdot T)$, we have $\ln(n) \sim T$. Thus, the dependence

$$e_n \sim \mu + \sigma \cdot \gamma \cdot \sqrt{2 \ln(n)}$$

of quality $q \stackrel{\text{def}}{=} e_n$ on time takes the form

$$q \approx a + b \cdot \sqrt{T}.$$

This is exactly the empirical dependence that we actually observe.

Thus, *the empirical data confirm the above theoretical formula.*

Comment. It is important to be cautious when testing the formula. For example, in a seemingly similar case of cars, the driving force for their fuel efficiency is not computer design but rather federal and state regulations which prescribe what fuel efficiency should be. Because of this, for cars, the dependence of fuel efficiency on time T is determined by the political will and is, thus, not as regular as for the aircraft.

Chapter 6
Decision Making Under Uncertainty and Restrictions on Computation Resources: From Heuristic to Optimal Techniques

In the previous chapters, we showed that bounded rationality explains the success of heuristic techniques in expert decision making. In this chapter, we explain how we can *improve* on the existing *heuristic techniques* by formulating and solving the corresponding optimization problems.

6.1 Decision Making Under Uncertainty: Towards Optimal Decisions

A natural idea of decision making under uncertainty is to assign a fair price to different alternatives, and then to use these fair prices to select the best alternative. In this section, we show how to assign a fair price under different types of uncertainty.

Results from this section first appeared in [74, 83, 84, 102].

6.1.1 Decision Making Under Uncertainty: Formulation of the Problem

In many practical situations, we have several alternatives, and we need to select one of these alternatives. For example:

- a person saving for retirement needs to find the best way to invest money;
- a company needs to select a location for its new plant;
- a designer must select one of several possible designs for a new airplane;
- a medical doctor needs to select a treatment for a patient, etc.

© Springer International Publishing AG 2018
J. Lorkowski and V. Kreinovich, *Bounded Rationality in Decision Making Under Uncertainty: Towards Optimal Granularity*, Studies in Systems, Decision and Control 99, DOI 10.1007/978-3-319-62214-9_6

Decision making is the easiest if we know the exact consequences of selecting each alternative. Often, however, we only have an incomplete information about consequences of different alternative, and we need to select an alternative under this uncertainty.

Traditional decision theory (see, e.g., [97, 134]) assumes that for each alternative a, we know the probability $p_i(a)$ of different outcomes i. It can be proven that preferences of a rational decision maker can be described by *utilities* u_i so that an alternative a is better if its expected utility $u(a) \stackrel{\text{def}}{=} \sum_i p_i(a) \cdot u_i$ is larger.

Often, we do not know the probabilities $p_i(a)$. As a result, we do not know the exact value of the gain u corresponding to each alternative. How can we then make a decision?

For the case when we only know the interval $[\underline{u}, \overline{u}]$ containing the actual (unknown) value of the gain u, a possible solution was proposed in the 1950s by a future Nobelist L. Hurwicz [56, 97]: we should select an alternative that maximizes the value $\alpha_H \cdot \overline{u}(a) + (1 - \alpha_H) \cdot \underline{u}(a)$. Here, the parameter $\alpha_H \in [0, 1]$ described the optimism level of a decision maker:

- $\alpha_H = 1$ means optimism;
- $\alpha_H = 0$ means pessimism;
- $0 < \alpha_H < 1$ combines optimism and pessimism.

Hurwicz's approach is widely used in decision making, but it is largely a heuristic, and it is not clear how to extend it other types of uncertainty. It is therefore desirable to develop more theoretically justified recommendations for decision making under uncertainty, recommendations that would be applicable to different types of uncertainty.

In this section, we propose such recommendations by explaining how to assign a fair price to each alternative, so that we can select between several alternatives by comparing their fair prices.

The structure of this section is as follows: first, we recall how to describe different types of uncertainty; then, we describe the fair price approach; after that, we show how the fair price approach can be applied to different types of uncertainty.

6.1.2   How to Describe Uncertainty

When we have a full information about a situation, then we can express our desirability of each possible alternative by declaring a price that we are willing to pay for this alternative. Once these prices are determined, we simply select the alternative for which the corresponding price is the highest. In this full information case, we know the exact gain u of selecting each alternative.

In practice, we usually only have partial information about the gain u: based on the available information, there are several possible values of the gain u. In other words, instead of the exact gain u, we only know a *set* S of possible values of the gain.

We usually know lower and upper bounds for this set, so this set is *bounded*. It is also reasonable to assume that the set S is *closed*: indeed, if we have a sequence of possible values $u_n \in S$ that converges to a number u_0, then, no matter how accurately we measure the gain, we can never distinguish between the limit value u_0 and a value u_n which is sufficiently close to this limit u_0. Thus, we will never be able to conclude that the limit value u_0 is not possible—and thus, it is reasonable to consider it possible, i.e., to include the limit point u_0 into the set S of possible values.

In many practical situations, if two gain values $u < u'$ are possible, then all intermediate values $u'' \in (u, u')$ are possible as well. In this case, the bounded closed set S is simply an *interval* $[\underline{u}, \overline{u}]$.

However, sometimes, some intermediate numbers u'' cannot be possible values of the gain. For example, if we buy an obscure lottery ticket for a simple prize-or-no-prize lottery from a remote country, we either get the prize or lose the money. In this case, the set of possible values of the gain consists of two values. To account for such situations, we need to consider general bounded closed sets.

In addition to knowing which gain values are possible, we may also have an information about which of these values are more probable and which values are less probable. Sometimes, this information has a *qualitative* nature, in the sense that, in addition to the set S of possible gain values, we also know a (closed) subset $s \subseteq S$ of values which are more probable (so that all the values from the difference $S - s$ are less probable). In many cases, the set s also contains all its intermediate values, so it is an interval; an important particular case is when this interval s consists of a single point. In other cases, the set s may be different from an interval.

Often, we have a *quantitative* information about the probability (frequency) of different values $u \in S$. A universal way to describe a probability distribution on the real line is to describe its cumulative distribution function (cdf) $F(u) \stackrel{\text{def}}{=} \text{Prob}(U \leq u)$. In the ideal case, we know the exact cdf $F(u)$. In practice, we usually only know the values of the cdf with uncertainty. Typically, for every u, we may only know the bounds $\underline{F}(u)$ and $\overline{F}(u)$ on the actual (unknown) values $F(u)$. The corresponding interval-valued function $[\underline{F}(u), \overline{F}(u)]$ is known as a *p-box* [32, 34].

The above description relates to the usual *passive* uncertainty, uncertainty over which we have no control. Sometimes, however, we have *active* uncertainty. As an example, let us consider two situations in which we need to minimize the amount of energy E used to heat the building. For simplicity, let us assume that cooling by 1 degree requires 1 unit of energy.

In the first situation, we simply know the interval $[\underline{E}, \overline{E}]$ that contains the actual (unknown) value of the energy E: for example, we know that $E \in [20, 25]$ (and we do not control this energy). In the second situation, we know that the outside temperature is between 50 and 55 F, and we want to maintain the temperature 75 F. In this case, we also conclude that $E \in [20, 25]$, but this time, we ourselves (or, alternatively, the heating system programmed by us) set up the appropriate amount of energy.

The distinction between the usual (passive) uncertainty and a different (active) type of uncertainty can be captured by considering *improper intervals* first introduced by Kaucher, i.e., intervals $[\underline{u}, \overline{u}]$ in which we may have $\underline{u} > \overline{u}$ see, e.g., [62, 142]. For example, in terms of these Kaucher intervals, our first (passive) situation is described by the interval $[15, 20]$, while the second (active) situation is described by an improper interval $[20, 15]$.

In line with this classification of different types of uncertainty, in the following text, we will first consider the simplest (interval) uncertainty, then the general set-valued uncertainty, then uncertainty described by a pair of embedded sets (in particular, by a pair of embedded intervals). After that, we consider situations with known probability distribution, situations with a known p-box, and finally, situations described by Kaucher intervals.

6.1.3   Fair Price Approach: Main Idea

When we have full information, we can express our desirability of each possible situation by declaring a price that we are willing to pay to get involved in this situation. To make decisions under uncertainty, it is therefore desirable to assign a fair price to each uncertain situation: e.g., to assign a fair price to each interval and/or to each set.

There are reasonable restrictions on the function that assigns the fair price to each type of uncertainty. First, the fair price should be *conservative*: if we know that the gain is always larger than or equal to $\underline{u}$, then the fair price corresponding to this situation should also be greater than or equal to $\underline{u}$. Similarly, if we know that the gain is always smaller than or equal to $\overline{u}$, then the fair price corresponding to this situation should also be smaller than or equal to $\overline{u}$.

Another natural property is *monotonicity*: if one alternative is clearly better than the other, then its fair price should be higher (or at least not lower).

Finally, the fair price should be *additive* in the following sense. Let us consider the situation when we have two consequent independent decisions. In this case, we can either consider two decision processes separately, or we can consider a single decision process in which we select a pair of alternatives:

- the 1st alternative corresponding to the 1st decision, and
- the 2nd alternative corresponding to the 2nd decision.

If we are willing to pay the amount u to participate in the first process, and we are willing to pay the amount v to participate in the second decision process, then it is reasonable to require that we should be willing to pay $u + v$ to participate in both decision processes.

On the examples of the above-mentioned types of uncertainty, let us describe the formulas for the fair price that can be derived from these requirements.

6.1.4 Case of Interval Uncertainty

We want to assign, to each interval $[\underline{u}, \overline{u}]$, a number $P([\underline{u}, \overline{u}])$ describing the fair price of this interval. Conservativeness means that the fair price $P([\underline{u}, \overline{u}])$ should be larger than or equal to $\underline{u}$ and smaller than or equal to $\overline{u}$, i.e., that the fair price of an interval should be located in this interval:

$$P([\underline{u}, \overline{u}]) \in [\underline{u}, \overline{u}].$$

Let us now apply monotonicity. Suppose that we keep the lower endpoint $\underline{u}$ intact but increase the upper bound. This means that we keep all the previous possibilities, but we also add new possibilities, with a higher gain. In other words, we are improving the situation. In this case, it is reasonable to require that after this addition, the fair price should either increase or remain the same, but it should definitely not decrease:

$$\text{if } \underline{u} = \underline{v} \text{ and } \overline{u} < \overline{v} \text{ then } P([\underline{u}, \overline{u}]) \leq P([\underline{v}, \overline{v}]).$$

Similarly, if we dismiss some low-gain alternatives, this should increase (or at least not decrease) the fair price:

$$\text{if } \underline{u} < \underline{v} \text{ and } \overline{u} = \overline{v} \text{ then } P([\underline{u}, \overline{u}]) \leq P([\underline{v}, \overline{v}]).$$

Finally, let us apply additivity. In the case of interval uncertainty, about the gain u from the first alternative, we only know that this (unknown) gain is in $[\underline{u}, \overline{u}]$. Similarly, about the gain v from the second alternative, we only know that this gain belongs to the interval $[\underline{v}, \overline{v}]$.

The overall gain $u + v$ can thus take any value from the interval

$$[\underline{u}, \overline{u}] + [\underline{v}, \overline{v}] \stackrel{\text{def}}{=} \{u + v : u \in [\underline{u}, \overline{u}], v \in [\underline{v}, \overline{v}]\}.$$

It is easy to check that (see, e.g., [60, 108]):

$$[\underline{u}, \overline{u}] + [\underline{v}, \overline{v}] = [\underline{u} + \underline{v}, \overline{u} + \overline{v}].$$

Thus, for the case of interval uncertainty, the additivity requirement about the fair prices takes the form

$$P([\underline{u} + \underline{v}, \overline{u} + \overline{v}]) = P([\underline{u}, \overline{u}]) + P([\underline{v}, \overline{v}]).$$

So, we arrive at the following definition:

Definition 6.1.1 By a *fair price under interval uncertainty*, we mean a function $P([\underline{u}, \overline{u}])$ for which:

- $\underline{u} \le P([\underline{u}, \overline{u}]) \le \overline{u}$ for all $\underline{u}$ and $\overline{u}$ (*conservativeness*);
- if $\underline{u} = \underline{v}$ and $\overline{u} < \overline{v}$, then $P([\underline{u}, \overline{u}]) \le P([\underline{v}, \overline{v}])$ (*monotonicity*);
- (*additivity*) for all $\underline{u}$, $\overline{u}$, $\underline{v}$, and $\overline{v}$, we have

$$P([\underline{u} + \underline{v}, \overline{u} + \overline{v}]) = P([\underline{u}, \overline{u}]) + P([\underline{v}, \overline{v}]).$$

Proposition 6.1.1 [102] *Each fair price under interval uncertainty has the form*

$$P([\underline{u}, \overline{u}]) = \alpha_H \cdot \overline{u} + (1 - \alpha_H) \cdot \underline{u} \text{ for some } \alpha_H \in [0, 1].$$

Comment. We thus get a new justification of Hurwicz optimism-pessimism criterion.

Proof

$1°$. Due to monotonicity, $P([u, u]) = u$.

$2°$. Also, due to monotonicity, $\alpha_H \overset{\text{def}}{=} P([0, 1]) \in [0, 1]$.

$3°$. For $[0, 1] = [0, 1/n] + \cdots + [0, 1/n]$ (n times), additivity implies $\alpha_H = n \cdot P([0, 1/n])$, so $P([0, 1/n]) = \alpha_H \cdot (1/n)$.

$4°$. For $[0, m/n] = [0, 1/n] + \cdots + [0, 1/n]$ (m times), additivity implies

$$P([0, m/n]) = \alpha_H \cdot (m/n).$$

$5°$. For each real number r, for each n, there is an m such that $m/n \le r \le (m + 1)/n$. Monotonicity implies

$$\alpha_H \cdot (m/n) = P([0, m/n]) \le P([0, r]) \le P([0, (m + 1)/n]) = \alpha_H \cdot ((m + 1)/n).$$

When $n \to \infty$, $\alpha_H \cdot (m/n) \to \alpha_H \cdot r$ and $\alpha_H \cdot ((m + 1)/n) \to \alpha_H \cdot r$, hence $P([0, r]) = \alpha_H \cdot r$.

$6°$. For $[\underline{u}, \overline{u}] = [\underline{u}, \underline{u}] + [0, \overline{u} - \underline{u}]$, additivity implies $P([\underline{u}, \overline{u}]) = \underline{u} + \alpha_H \cdot (\overline{u} - \underline{u})$. The proposition is proven.

6.1.5 Case of Set-Valued Uncertainty

Intervals are a specific case of bounded closed sets. We already know how to assign fair price to intervals. So, we arrive at the following definition.

Definition 6.1.2 By a *fair price under set-valued uncertainty*, we mean a function P that assigns, to every bounded closed set S, a real number $P(S)$, for which:

- $P([\underline{u}, \overline{u}]) = \alpha_H \cdot \overline{u} + (1 - \alpha_H) \cdot \underline{u}$ (*conservativeness*);
- $P(S + S') = P(S) + P(S')$, where $S + S' \overset{\text{def}}{=} \{s + s' : s \in S, s' \in S'\}$ (*additivity*).

Proposition 6.1.2 *Each fair price under set uncertainty has the form* $P(S) = \alpha_H \cdot \sup S + (1 - \alpha_H) \cdot \inf S$.

Proof It is easy to check that each bounded closed set S contains its infimum $\underline{S} \overset{\text{def}}{=} \inf S$ and supremum $\underline{S} \overset{\text{def}}{=} \sup S$: $\{\underline{S}, \overline{S}\} \subseteq S \subseteq [\underline{S}, \overline{S}]$. Thus,

$$[2\underline{S}, 2\overline{S}] = \{\underline{S}, \overline{S}\} + [\underline{S}, \overline{S}] \subseteq S + [\underline{S}, \overline{S}] \subseteq [\underline{S}, \overline{S}] + [\underline{S}, \overline{S}] = [2\underline{S}, 2\overline{S}].$$

So, $S + [\underline{S}, \overline{S}] = [2\underline{S}, 2\overline{S}]$. By additivity, we conclude that $P(S) + P([\underline{S}, \overline{S}]) = P([2\underline{S}, 2\overline{S}])$. Due to conservativeness, we know the fair prices $P([\underline{S}, \overline{S}])$ and $P([2\underline{S}, 2\overline{S}])$. Thus, we can conclude that

$$P(S) = P([2\underline{S}, 2\overline{S}]) - P([\underline{S}, \overline{S}]) = (\alpha_H \cdot (2\overline{S}) + (1 - \alpha_H) \cdot (2\underline{S})) - (\alpha_H \cdot \overline{S} + (1 - \alpha_H) \cdot \underline{S}),$$

hence indeed $P(S) = \alpha_H \cdot \overline{S} + (1 - \alpha_H) \cdot \underline{S}$. The proposition is proven.

6.1.6 Case of Embedded Sets

In addition to a set S of possible values of the gain u, we may also know a subset $s \subseteq S$ of more probable values u. To describe a fair price assigned to such a pair (S, s), let us start with the simplest case when the original set S is an interval $S = [\underline{u}, \overline{u}]$, and the subset s is a single "most probable" value u_0 within this interval. Such pairs are known as *triples*; see, e.g., [22] and references therein. For triples, addition is defined component-wise:

$$([\underline{u}, \overline{u}], u_0) + ([\underline{v}, \overline{v}], v_0) = ([\underline{u} + \underline{v}, \overline{u} + \overline{v}], u_0 + v_0).$$

Thus, the additivity requirement about the fair prices takes the form

$$P([\underline{u} + \underline{v}, \overline{u} + \overline{v}], u_0 + v_0) = P([\underline{u}, \overline{u}], u_0) + P([\underline{v}, \overline{v}], v_0).$$

Definition 6.1.3 By a *fair price under triple uncertainty*, we mean a function $P([\underline{u}, \overline{u}], u_0)$ for which:

- $\underline{u} \leq P([\underline{u}, \overline{u}], u_0) \leq \overline{u}$ for all $\underline{u} \leq u \leq \overline{u}$ (*conservativeness*);
- if $\underline{u} \leq \underline{v}$, $u_0 \leq v_0$, and $\overline{u} \leq \overline{v}$, then $P([\underline{u}, \overline{u}], u_0) \leq P([\underline{v}, \overline{v}], v_0)$ (*monotonicity*);
- (*additivity*) for all $\underline{u}$, $\overline{u}$, u_0 $\underline{v}$, $\overline{v}$, and v_0, we have

$$P([\underline{u} + \underline{v}, \overline{u} + \overline{v}], u_0 + v_0) = P([\underline{u}, \overline{u}], u_0) + P([\underline{v}, \overline{v}], v_0).$$

Proposition 6.1.3 *Each fair price under triple uncertainty has the form*

$$P([\underline{u}, \overline{u}], u_0) = \alpha_L \cdot \underline{u} + (1 - \alpha_L - \alpha_U) \cdot u_0 + \alpha_U \cdot \overline{u}, \text{ where } \alpha_L, \alpha_U \in [0, 1].$$

Proof In general, we have

$$([\underline{u}, \overline{u}], u_0) = ([u_0, u_0], u_0) + ([0, \overline{u} - u], 0) + ([\underline{u} - u, 0], 0).$$

So, due to additivity:

$$P([\underline{u}, \overline{u}], u_0) = P([u_0, u_0], u_0) + P([0, \overline{u} - u_0], 0) + P([\underline{u} - u_0, 0], 0).$$

Due to conservativeness, $P([u_0, u_0], u_0) = u_0$.

Similarly to the interval case, we can prove that $P([0, r], 0) = \alpha_U \cdot r$ for some $\alpha_U \in [0, 1]$, and that $P([r, 0], 0) = \alpha_L \cdot r$ for some $\alpha_L \in [0, 1]$. Thus,

$$P([\underline{u}, \overline{u}], u_0) = \alpha_L \cdot \underline{u} + (1 - \alpha_L - \alpha_U) \cdot u_0 + \alpha_U \cdot \overline{u}.$$

The proposition is proven.

The next simplest case is when both sets S and $s \subseteq S$ are intervals, i.e., when, inside the interval $S = [\underline{u}, \overline{u}]$, instead of a "most probable" value u_0, we have a "most probable" subinterval $[\underline{m}, \overline{m}] \subseteq [\underline{u}, \overline{u}]$. The resulting pair of intervals is known as a "twin interval" (see, e.g., [41, 113]).

For such twin intervals, addition is defined component-wise:

$$([\underline{u}, \overline{u}], [\underline{m}, \overline{m}]) + ([\underline{v}, \overline{v}], [\underline{n}, \overline{n}]) = ([\underline{u} + \underline{v}, \overline{u} + \overline{v}], [\underline{m} + \underline{n}, \overline{m} + \overline{n}]).$$

Thus, the additivity requirement about the fair prices takes the form

$$P([\underline{u} + \underline{v}, \overline{u} + \overline{v}], [\underline{m} + \underline{n}, \overline{m} + \overline{n}]) = P([\underline{u}, \overline{u}], [\underline{m}, \overline{m}]) + P([\underline{v}, \overline{v}], [\underline{n}, \overline{n}]).$$

Definition 6.1.4 By a *fair price under twin uncertainty*, we mean a function $P([\underline{u}, \overline{u}], [\underline{m}, \overline{m}])$ for which:

- $\underline{u} \leq P([\underline{u}, \overline{u}], [\underline{m}, \overline{m}]) \leq \overline{u}$ for all $\underline{u} \leq \underline{m} \leq \overline{m} \leq \overline{u}$ (*conservativeness*);
- if $\underline{u} \leq \underline{v}$, $\underline{m} \leq \underline{n}$, $\overline{m} \leq \overline{n}$, and $\overline{u} \leq \overline{v}$, then $P([\underline{u}, \overline{u}], [\underline{m}, \overline{m}]) \leq P([\underline{v}, \overline{v}], [\underline{n}, \overline{n}])$ (*monotonicity*);
- for all $\underline{u} \leq \underline{m} \leq \overline{m} \leq \overline{u}$ and $\underline{v} \leq \underline{n} \leq \overline{n} \leq \overline{v}$, we have *additivity*:

$$P([\underline{u} + \underline{v}, \overline{u} + \overline{v}], [\underline{m} + \underline{n}, \overline{m} + \overline{m}]) = P([\underline{u}, \overline{u}], [\underline{m}, \overline{m}]) + P([\underline{v}, \overline{v}], [\underline{n}, \overline{n}]).$$

Proposition 6.1.4 *Each fair price under twin uncertainty has the following form, for some $\alpha_L, \alpha_u, \alpha_U \in [0, 1]$:*

$$P([\underline{u}, \overline{u}], [\underline{m}, \overline{m}]) = \underline{m} + \alpha_u \cdot (\overline{m} - \underline{m}) + \alpha_U \cdot (\overline{u} - \overline{m}) + \alpha_L \cdot (\underline{u} - \underline{m}).$$

Proof In general, we have

$$([\underline{u}, \overline{u}], [\underline{m}, \overline{m}]) = ([\underline{m}, \underline{m}], [\underline{m}, \underline{m}]) + ([0, \overline{m} - \underline{m}], [0, \overline{m} - \underline{m}]) +$$

$$([0, \overline{u} - \overline{m}], [0, 0]) + ([\underline{u} - \underline{m}, 0], [0, 0])].$$

So, due to additivity:

$$P([\underline{u}, \overline{u}], [\underline{m}, \overline{m}]) = P([\underline{m}, \underline{m}], [\underline{m}, \underline{m}]) + P([0, \overline{m} - \underline{m}], [0, \overline{m} - \underline{m}]) +$$

$$P([0, \overline{u} - \overline{m}], [0, 0]) + P([\underline{u} - \underline{m}, 0], [0, 0])].$$

Due to conservativeness, $P([\underline{m}, \underline{m}], [\underline{m}, \underline{m}]) = \underline{m}$. Similarly to the interval case, we can prove that:

- $P([0, r], [0, r]) = \alpha_u \cdot r$ for some $\alpha_u \in [0, 1]$,
- $P([0, r], [0, 0]) = \alpha_U \cdot r$ for some $\alpha_U \in [0, 1]$;
- $P([r, 0], [0, 0]) = \alpha_L \cdot r$ for some $\alpha_L \in [0, 1]$.

Thus,

$$P([\underline{u}, \overline{u}], [\underline{m}, \overline{m}]) = \underline{m} + \alpha_u \cdot (\overline{m} - \underline{m}) + \alpha_U \cdot (\overline{u} - \overline{m}) + \alpha_L \cdot (\underline{u} - \underline{m}).$$

The proposition is proven.

Finally, let us consider the general case.

Definition 6.1.5 By a *fair price under embedded-set uncertainty*, we mean a function P that assigns, to every pair of bounded closed sets (S, s) with $s \subseteq S$, a real number $P(S, s)$, for which:

- $P([\underline{u}, \overline{u}], [\underline{m}, \overline{m}]) = \underline{m} + \alpha_u \cdot (\overline{m} - \underline{m}) + \alpha_U \cdot (\overline{U} - \overline{m}) + \alpha_L \cdot (\underline{u} - \underline{m})$
  (*conservativeness*);
- $P(S + S', s + s') = P(S, s) + P(S', s')$ (*additivity*).

Proposition 6.1.5 *Each fair price under embedded-set uncertainty has the form*

$$P(S, s) = \inf s + \alpha_u \cdot (\sup s - \inf s) + \alpha_U \cdot (\sup S - \sup s) + \alpha_L \cdot (\inf S - \inf s).$$

Proof Similarly to the proof of Proposition 6.1.2, we can conclude that

$$(S, s) + ([\inf S, \sup S], [\inf s, \sup s]) = ([2 \cdot \inf S, 2 \cdot \sup S], [2 \cdot \inf s, 2 \cdot \sup s]).$$

By additivity, we conclude that

$$P(S, s) + P([\inf S, \sup S], [\inf s, \sup s]) =$$

$$P([2 \cdot \inf S, 2 \cdot \sup S], [2 \cdot \inf s, 2 \cdot \sup s]),$$

hence

$$P(S, s) = P([2 \cdot \inf S, \cdot \sup S], [2 \cdot \inf s, 2 \cdot \sup s]) -$$

$$P([\inf S, \sup S], [\inf s, \sup s]).$$

Due to conservativeness, we know the fair prices

$$P([2 \cdot \inf S, 2 \cdot \sup S], [2 \cdot \inf s, 2 \cdot \sup s]) \text{ and } P([\inf S, \sup S], [\inf s, \sup s]).$$

Subtracting these expressions, we get the desired formula for $P(S, s)$. The proposition is proven.

6.1.7 Cases of Probabilistic and P-Box Uncertainty

Suppose that for some financial instrument, we know the corresponding probability distribution $F(u)$ on the set of possible gains u. What is the fair price P for this instrument?

Due to additivity, the fair price for n copies of this instrument is $n \cdot P$. According to the Large Numbers Theorem, for large n, the average gain tends to the mean value $\mu = \int u \, dF(u)$.

Thus, the fair price for n copies of the instrument is close to $n \cdot \mu$: $n \cdot P \approx n \cdot \mu$. The larger n, the closer the averages. So, in the limit, we get $P = \mu$.

So, the fair price under probabilistic uncertainty is equal to the average gain $\mu = \int u \, dF(u)$.

Let us now consider the case of a p-box $[\underline{F}(u), \overline{F}(u)]$. For different functions $F(u) \in [\underline{F}(u), \overline{F}(u)]$, values of the mean μ form an interval $\left[\underline{\mu}, \overline{\mu}\right]$, where $\underline{\mu} = \int u \, d\overline{F}(u)$ and $\overline{\mu} = \int u \, d\underline{F}(u)$. Thus, the price of a p-box is equal to the price of an interval $\left[\underline{\mu}, \overline{\mu}\right]$.

We already know that the fair price of this interval is equal to

$$\alpha_H \cdot \overline{\mu} + (1 - \alpha_H) \cdot \underline{\mu}.$$

Thus, we conclude that the fair price of a p-box $[\underline{F}(u), \overline{F}(u)]$ is $\alpha_H \cdot \overline{\mu} + (1 - \alpha_H) \cdot \underline{\mu}$, where $\underline{\mu} = \int u \, d\overline{F}(u)$ and $\overline{\mu} = \int u \, d\underline{F}(u)$.

6.1.8 Case of Kaucher (Improper) Intervals

For Kaucher intervals, addition is also defined component-wise; in particular, for all $\underline{u} < \overline{u}$, we have

$$[\underline{u}, \overline{u}] + [\overline{u}, \underline{u}] = [\underline{u} + \overline{u}, \underline{u} + \overline{u}].$$

Thus, additivity implies that

$$P([\underline{u}, \overline{u}]) + P([\overline{u}, \underline{u}]) = P([\underline{u} + \overline{u}, \underline{u} + \overline{u}]).$$

We know that $P([\overline{u}, \underline{u}]) = \alpha_H \cdot \underline{u} + (1 - \alpha_H) \cdot \overline{u}$ and $P([\underline{u} + \overline{u}, \underline{u} + \overline{u}]) = \underline{u} + \overline{u}$. Hence:

$$P([\underline{u}, \overline{u}]) = (\underline{u} + \overline{u}) - (\alpha_H \cdot \underline{u} + (1 - \alpha_H) \cdot \overline{u}).$$

Thus, the fair price $P([\underline{u}, \overline{u}])$ of an improper interval $[\underline{u}, \overline{u}]$, with $\underline{u} > \overline{u}$, is equal to $P([\underline{u}, \overline{u}]) = \alpha_H \cdot \overline{u} + (1 - \alpha_H) \cdot \underline{u}$.

6.1.9 (Crisp) Z-Numbers, Z-Intervals, and Z-Sets: Cases When the Probabilities Are Crisp

In the previous sections, we assumed that we are 100% certain that the actual gain is contained in the given interval (or set). In reality, mistakes are possible, so usually, we are only certain that u belongs to the corresponding interval or set with some probability $0 < p < 1$. In such situations, to fully describe our knowledge, we need to describe both the interval (or set) *and* this probability p.

In the general context, after supplementing the information about a quantity with the information of how certain we are about this piece of information, we get what L. Zadeh calls a *Z-number* [168]. Because of this:

- we will call a pair consisting of a (crisp) number and a (crisp) probability a *crisp Z-number*;
- we will call a pair consisting of an interval and a probability a *Z-interval*; and
- we will call a pair consisting of a set and a probability a *Z-set*.

In this section, we will describe fair prices for crisp Z-numbers, Z-intervals, and Z-sets for situations when the probability p is known exactly.

How can we define operations on Z-numbers? When we have two independent sequential decisions, and we are 100% sure that the first decision leads to gain u and the second decision leads to gain v, then, as we have mentioned earlier, the user's total gain is equal to the sum $u + v$. In this section, we consider the situation in which:

- for the first decision, our degree of confidence in the gain estimate u is described by some probability p;
- for the second decision, our degree of confidence in the gain estimate v is described by some probability q.

The estimate $u + v$ is valid only if both gain estimates are correct. Since these estimates are independent, the probability that they are both correct is equal to the product $p \cdot q$ of the corresponding probabilities. Thus:

- for crisp Z-numbers (u, p) and (v, q), the sum is equal to $(u + v, p \cdot q)$;
- for Z-intervals $([\underline{u}, \overline{u}], p)$ and $[\underline{v}, \overline{v}], q)$, the sum is equal to $([\underline{u} + \underline{v}, \overline{u} + \overline{v}], p \cdot q)$;
- finally, for Z-sets (S, p) and (S', q), the sum is equal to $(S + S', p \cdot q)$.

Let us analyze these cases one by one.

Let us start with the case of crisp Z-numbers. Since the probability p is usually known with some uncertainty, it makes sense to require that the fair price of a crisp Z-number (u, p) continuously depend on p, so that small changes in p lead to small changes in the fair price—and the closer our estimate to the actual value of the probability, the closer the estimated fair price should be to the actual fair price.

Thus, we arrive at the following definitions.

Definition 6.1.6 By a *crisp Z-number*, we mean a pair (u, p) of two real numbers such that $0 < p \le 1$.

Definition 6.1.7 By a *fair price under crisp Z-number uncertainty*, we mean a function $P(u, p)$ that assigns, to every crisp Z-number, a real number, and which satisfies the following properties:

- $P(u, 1) = u$ for all u (conservativeness);
- for all u, v, p, and q, we have $P(u + v, p \cdot q) = P(u, p) + P(v, q)$ (additivity);
- the function $P(u, p)$ is continuous in p (continuity).

Proposition 6.1.6 *Each fair price under crisp Z-number uncertainty has the form* $P(u, p) = u - k \cdot \ln(p)$ *for some real number k.*

Proof

$1°$. By additivity, we have $P(u, p) = P(u, 1) + P(0, p)$. By conservativeness, we have $P(u, 1) = u$; thus, $P(u, p) = u + P(0, p)$. So, it is clear that to find the general expression for the fair price function $P(u, p)$, it is sufficient to find the values $P(0, p)$ corresponding to $u = 0$.

$2°$. Additivity implies that $P(0, p \cdot q) = P(0, p) + P(0, q)$.

$3°$. Let us first consider the value $p = e^{-1}$ which corresponds to $\ln(p) = -1$. The corresponding value of $P(0, p)$ will be denoted by $k \overset{\text{def}}{=} P(0, e^{-1})$. Then, for $p = e^{-1}$, we have the desired expression $P(0, p) = -k \cdot \ln(p)$.

$4°$. Let us now consider the values $P(0, e^{-m})$ for positive integer values m. The probability e^{-m} can be represented as a product of m values e^{-1}:

$$e^{-m} = e^{-1} \cdot \cdots \cdot e^{-1} \quad (m \text{ times}). \tag{6.1.1}$$

Thus, due to additivity, we have

$$P(0, e^{-m}) = P(0, e^{-1}) + \cdots + P(0, e^{-1}) \quad (m \text{ times}) = m \cdot k. \tag{6.1.2}$$

Since for $p = e^{-m}$, we have $\ln(p) = -m$, we thus have $P(0, p) = -k \cdot \ln(p)$ for these values p.

$5°$. Now, let us estimate the value $P(0, p)$ for $p = e^{-1/n}$, for a positive integer n.

In this case, the value e^{-1} can be represented as a product of n probabilities equal to $e^{-1/n}$:

$$e^{-1} = e^{-1/n} \cdot \cdots \cdot e^{-1/n} \quad (n \text{ times}).$$

Thus, due to additivity, we have

$$k = P(0, e^{-1}) = P(0, e^{-1/n}) + \cdots + P(0, e^{-1/n}) \quad (n \text{ times}),$$

i.e.,

$$k = n \cdot P(0, e^{-1/n}) \tag{6.1.3}$$

and hence, $P(0, e^{-1/n}) = \dfrac{k}{n}$. Therefore, for $p = e^{-1/n}$, we also have $P(0, p) = -k \cdot \ln(p)$.

$6°$. For every two positive numbers $m > 0$ and $n > 0$, the probability $e^{-m/n}$ can be represented as the product of m probabilities equal to $e^{-1/n}$. Thus, due to additivity, we have $P(0, e^{-m/n}) = m \cdot P(0, e^{-1/n}) = k \cdot \dfrac{m}{n}$. Hence, for the values $p = e^{-m/n}$ for which the logarithm $\ln(p)$ is a rational number, we have $P(0, p) = -k \cdot \ln(p)$.

$7°$. Every real number $\ell \overset{\text{def}}{=} \ln(p)$ can be approximated, with arbitrary accuracy, by rational numbers $\ell_n \to \ell$ for which $p_n \overset{\text{def}}{=} e^{-\ell_n} \to e^{-\ell} = p$. For these rational numbers, we have $P(0, p_n) = -k \cdot \ln(p_n)$. Thus, when $n \to \infty$ and $p_n \to p$, by continuity, we have $P(0, p) = -k \cdot \ln(p)$.

From Part 1, we know that $P(u, p) = u + P(0, p)$; thus, indeed, $P(u, p) = u - k \cdot \ln(p)$. The proposition is proven.

Similar results hold for Z-intervals and Z-sets; in both results, we will use the fact that we already know how to set a fair price for the case when $p = 1$.

Definition 6.1.8 By a Z-*interval*, we mean a pair $([\underline{u}, \overline{u}], p)$ consisting of an interval $[\underline{u}, \overline{u}]$ and a real number p such that $0 < p \leq 1$.

Definition 6.1.9 By a *fair price under Z-interval uncertainty*, we mean a function $P([\underline{u}, \overline{u}], p)$ that assigns, to every Z-interval, a real number, and which satisfies the following properties:

- for some $\alpha_H \in [0, 1]$ and for all $\underline{u} \leq \overline{u}$, we have $P([\underline{u}, \overline{u}], 1) = \alpha_H \cdot \overline{u} + (1 - \alpha_H) \cdot \underline{u}$ (conservativeness);
- for all $\underline{u}, \overline{u}, \underline{v}, \overline{v}, p$, and q, we have

$$P([\underline{u} + \underline{v}, \overline{u} + \overline{v}], p \cdot q) = P([\underline{u}, \overline{u}], p) + P([\underline{v}, \overline{v}], q) \tag{6.1.4}$$

(additivity).

Proposition 6.1.7 *Each fair price under Z-interval uncertainty has the form* $P([\underline{u}, \overline{u}], p) = \alpha_H \cdot \overline{u} + (1 - \alpha_H) \cdot \underline{u} - k \cdot \ln(p)$ *for some real numbers* $\alpha_H \in [0, 1]$ *and* k.

Proof By additivity, we have $P([\underline{u}, \overline{u}], p) = P([\underline{u}, \overline{u}], 1) + P(0, p)$. By conservativeness, we have

$$P([\underline{u}, \overline{u}], 1) = \alpha_H \cdot \overline{u} + (1 - \alpha_H) \cdot \underline{u}. \tag{6.1.5}$$

For $P(0, p)$, similarly to the proof of Proposition 6.1.6, we conclude that $P(0, p) = -k \cdot \ln(p)$ for some real number k. The proposition is proven.

Definition 6.1.10 By a *Z-set*, we mean a pair (S, p) consisting of a closed bounded set S and a real number p such that $0 < p \leq 1$.

Definition 6.1.11 By a *fair price under Z-set-valued uncertainty*, we mean a function $P(S, p)$ that assigns, to every Z-set (S, p), a real number, and which satisfies the following properties:

- for some $\alpha_H \in [0, 1]$ and for all sets S, we have

$$P(S, 1) = \alpha_H \cdot \sup S + (1 - \alpha_H) \cdot \inf S \qquad (6.1.6)$$

(conservativeness);
- for all $S, S', p,$ and q, we have $P(S + S', p \cdot q) = P(S, p) + P(S', q)$ (additivity).

Proposition 6.1.8 *Each fair price under Z-set-valued uncertainty has the form*

$$P(S, p) = \alpha_H \cdot \sup S + (1 - \alpha_H) \cdot \inf S - k \cdot \ln(p) \qquad (6.1.7)$$

for some real numbers $\alpha_H \in [0, 1]$ and k.

Proof By additivity, we have $P(S, p) = P(S, 1) + P(\{0\}, p)$. By conservativeness, we have
$$P(S, 1) = \alpha_H \cdot \sup S + (1 - \alpha_H) \cdot \inf S.$$

For $P(\{0\}, p)$, similarly to the proof of Proposition 6.1.6, we conclude that $P(\{0\}, p) = -k \cdot \ln(p)$ for some real number k. The proposition is proven.

6.1.10 (Crisp) Z-Numbers, Z-Intervals, and Z-Sets: Cases When Probabilities Are Known with Interval or Set-Valued Uncertainty

When we know the exact probabilities p and q that the corresponding estimates are correct, then the probability that both estimates are correct is equal to the product $p \cdot q$.

Similarly to the fact that we often do not know the exact gain, we often do not know the exact probability p. Instead, we may only know the interval $\left[\underline{p}, \overline{p} \right]$ of possible values of p, or, more generally, a set $\mathscr{P}$ of possible values of p. If we know p and q with such uncertainty, what can we then conclude about the product $p \cdot q$?

For positive values p and q, the function $p \cdot q$ is increasing as a function of both variables: if we increase p and/or increase q, the product increases. Thus, if the only information that we have the probability p is that this probability belongs to the interval $[\underline{p}, \overline{p}]$, and the only information that we have the probability q is that this probability belongs to the interval $[\underline{q}, \overline{q}]$, then:

- the smallest possible value of $p \cdot q$ is equal to the product $\underline{p} \cdot \underline{q}$ of the smallest values;
- the largest possible value of $p \cdot q$ is equal to the product $\overline{p} \cdot \overline{q}$ of the largest values; and
- the set of all possible values $p \cdot q$ is the interval

$$\left[\underline{p} \cdot \underline{q}, \overline{p} \cdot \overline{q} \right].$$

For sets $\mathscr{P}$ and $\mathscr{Q}$, the set of possible values $p \cdot q$ is the set

$$\mathscr{P} \cdot \mathscr{Q} \overset{\text{def}}{=} \{p \cdot q : p \in \mathscr{P} \text{ and } q \in \mathscr{Q}\}. \tag{6.1.8}$$

Let us find the fair price under such uncertainty.

Let us start with the case of crisp Z-numbers under such uncertainty.

Definition 6.1.12 By a *crisp Z-number under interval p-uncertainty*, we mean a pair $(u, [\underline{p}, \overline{p}])$ consisting of a real number u and an interval $\left[\underline{p}, \overline{p} \right] \subseteq (0, 1]$.

Definition 6.1.13 By a *fair price under crisp Z-number p-interval uncertainty*, we mean a function $P \left(u, \left[\underline{p}, \overline{p} \right] \right)$ that assigns, to every crisp Z-number under interval p-uncertainty, a real number, and which satisfies the following properties:

- for some real number k, we have

$$P(u, [p, p]) = u - k \cdot \ln(p) \tag{6.1.9}$$

for all u and p (conservativeness);
- for all $u, v, \underline{p}, \overline{p}, \underline{q}$, and $\overline{q}$, we have

$$P \left(u + v, \left[\underline{p} \cdot \underline{q}, \overline{p} \cdot \overline{q} \right] \right) = P \left(u, \left[\underline{p}, \overline{p} \right] \right) + P \left(v, \left[\underline{q}, \overline{q} \right] \right) \tag{6.1.10}$$

(additivity);
- the function $P \left(u, \left[\underline{p}, \overline{p} \right] \right)$ is continuous in $\underline{p}$ and $\overline{p}$ (continuity).

Proposition 6.1.9 *Each fair price under crisp Z-number p-interval uncertainty has the form*

$$P \left(u, \left[\underline{p}, \overline{p} \right] \right) = u - (k - \beta) \cdot \ln(\overline{p}) - \beta \cdot \ln \left(\underline{p} \right) \tag{6.1.11}$$

for some real numbers k and $\beta \in [0, 1]$.

Proof

1°. By additivity, we have $P\left(u,\left[\underline{p},\overline{p}\right]\right) = P(u,\overline{p}) + P(0,[p,1])$, where $p \overset{\text{def}}{=}$ $\underline{p}/\overline{p}$. By conservativeness, we know that $P(u,\overline{p}) = u - k \cdot \ln(\overline{p})$. Thus, $P(u,p) = u - k \cdot \ln(\overline{p}) + P(0,[p,1])$. So, to find the general expression for the fair price function $P\left(u,\left[\underline{p},\overline{p}\right]\right)$, it is sufficient to find the values $P(0,[p,1])$ corresponding to $u = 0$ and $\overline{p} = 1$.

2°. For the values $P(0,[p,1])$, additivity implies that

$$P(0,[p \cdot q, 1]) = P(0,[p,1]) + P(0,[q,1]). \tag{6.1.12}$$

In Part 2 of the proof of Proposition 6.1.6, we had a similar property for a continuous function $P(0,p)$, and we proved, in Parts 2–6 of that proof, that this property implies that this continuous function is equal to $-c \cdot \ln(p)$ for some real number c. Thus, we can similarly conclude that

$$P(0,[p,1]) = \beta \cdot \ln(p) \tag{6.1.13}$$

for some real number $\beta = -c$.

3°. From Part 1 of this proof, we know that

$$P\left(u,\left[\underline{p},\overline{p}\right]\right) = u - k \cdot \ln(\overline{p}) + P(0,[p,1]); \tag{6.1.14}$$

thus,

$$P\left(u,\left[\underline{p},\overline{p}\right]\right) = u - k \cdot \ln(\overline{p}) + \beta \cdot \ln(p). \tag{6.1.15}$$

Substituting $p = \underline{p}/\overline{p}$ into this formula and taking into account that $\ln(p) = \ln(\underline{p}) - \ln(\overline{p})$, we get the desired formula.

Definition 6.1.14 By a *crisp Z-number under set-valued p-uncertainty*, we mean a pair $(u,\mathcal{P})$ consisting of a real number u and a bounded closed set $\mathcal{P} \subseteq (0,1]$.

Comment. One can easily show that for each closed set $\mathcal{P} \subseteq (0,1]$, we have $\inf \mathcal{P} > 0$.

Definition 6.1.15 By a *fair price under crisp Z-number p-set-valued uncertainty*, we mean a function $P(u,\mathcal{P})$ that assigns, to every crisp Z-number under set-valued p-uncertainty, a real number, and which satisfies the following properties:

• for some real numbers k and β, we have

$$P\left(u,\left[\underline{p},\overline{p}\right]\right) = u - (k - \beta) \cdot \ln(\overline{p}) - \beta \cdot \ln\left(\underline{p}\right) \tag{6.1.16}$$

for all u, $\underline{p}$, and $\overline{p}$ (conservativeness);

- for all u, v, $\mathscr{P}$, and $\mathscr{Q}$, we have

$$P(u + v, \mathscr{P} \cdot \mathscr{Q}) = P(u, \mathscr{P}) + P(v, \mathscr{Q}) \qquad (6.1.17)$$

(additivity).

Proposition 6.1.10 *Each fair price under crisp Z-number p-set-valued uncertainty has the form*

$$P(u, \mathscr{P}) = u - (k - \beta) \cdot \ln(\sup \mathscr{P}) - \beta \cdot \ln(\inf \mathscr{P}) \qquad (6.1.18)$$

for some real number $\beta \in [0, 1]$.

Proof By additivity, we have $P(u, \mathscr{P}) = P(u, \{1\}) + P(0, \mathscr{P})$, i.e., due to conservativeness, $P(u, \mathscr{P}) = u + P(0, \mathscr{P})$. So, to find the expression for $P(u, \mathscr{P})$, it is sufficient to find the values $P(0, \mathscr{P})$. Similarly to prove of Proposition 6.1.2, we can prove that

$$\mathscr{P} \cdot [\inf \mathscr{P}, \sup \mathscr{P}] = \left[(\inf \mathscr{P})^2, (\sup \mathscr{P})^2\right]. \qquad (6.1.19)$$

Due to additivity, this implies that

$$P\left(0, \left[(\inf \mathscr{P})^2, (\sup \mathscr{P})^2\right]\right) = P(0, \mathscr{P}) + P(0, [\inf \mathscr{P}, \sup \mathscr{P}]), \qquad (6.1.20)$$

hence

$$P(0, \mathscr{P}) = P\left(0, \left[(\inf \mathscr{P})^2, (\sup \mathscr{P})^2\right]\right) - P(0, [\inf \mathscr{P}, \sup \mathscr{P}]). \qquad (6.1.21)$$

Due to conservativeness, we know the values in the right-hand side of this equality. Substituting these values, we get the desired formula.

Let us extend the above results to Z-sets (and to their particular case: Z-intervals).

Definition 6.1.16 By a *Z-set under set-valued p-uncertainty*, we mean a pair $(S, \mathscr{P})$ consisting of a bounded closed set S and a bounded closed set $\mathscr{P} \subseteq (0, 1]$.

Definition 6.1.17 By a *fair price under Z-set p-set-valued uncertainty*, we mean a function $P(S, \mathscr{P})$ that assigns, to every Z-set under set-valued p-uncertainty, a real number, and which satisfies the following properties:

- for some real number $\alpha_H \in [0, 1]$, we have

$$P(S, 1) = \alpha_H \cdot \sup S + (1 - \alpha_H) \cdot \inf S \qquad (6.1.22)$$

for all S (conservativeness);

- for some real numbers k and β, we have

$$P(u, \mathscr{P}) = u - (k - \beta) \cdot \ln(\sup \mathscr{P}) - \beta \cdot \ln(\inf \mathscr{P}) \qquad (6.1.23)$$

for all u and $\mathscr{P}$ (conservativeness);
- for all S, S', $\mathscr{P}$, and $\mathscr{Q}$, we have

$$P(S + S', \mathscr{P} \cdot \mathscr{Q}) = P(S, \mathscr{P}) + P(Q, \mathscr{Q}) \qquad (6.1.24)$$

(additivity).

Proposition 6.1.11 *Each fair price under Z-set p-set-valued uncertainty has the form*

$$P(S, \mathscr{P}) = \alpha_H \cdot \sup S + (1 - \alpha_H) \cdot \inf S - (k - \beta) \cdot \ln(\overline{p}) - \beta \cdot \ln\left(\underline{p}\right). \qquad (6.1.25)$$

6.1.11 Case of Fuzzy Uncertainty

In the above text, we first considered situations when about each value of gain u, the expert is either absolutely sure that this value is possible or absolutely sure that this value is not possible. Then, we took into account the possibility that the expert is not 100% certain about that—but we assumed that the expert either knows the exact probability p describing his/her degree of certainty, or that the expert is absolutely sure which probabilities can describe his/her uncertainty and which cannot.

In reality, an expert is often uncertain about the possible values, and uncertain about possible degrees of uncertainty. To take this uncertainty into account, L. Zadeh introduced the notion of a *fuzzy set* [65, 120, 163], where, to each possible value of u, we assign a degree $\mu(u) \in [0, 1]$ to which this value u is possible. Similarly, a fuzzy set $\mu_p : [0, 1] \to [0, 1]$ can describe the degrees to which different probability values are possible.

In this section, we restrict ourselves to *fuzzy numbers s*, i.e., fuzzy sets for which the membership function is different from 0 only on a bounded set, where it first monotonically increases until it reaches a point $\overline{s}$ at which $\mu(\overline{s}) = 1$, and then monotonically decreases from 1–0.

Operations on fuzzy numbers are usually described in terms of *Zadeh's extension principle:* if two quantities u and v are described by membership functions $\mu_1(u)$ and $\mu_2(v)$, then their sum $w = u + v$ is described by the membership function $\mu(w) = \max_{u,v: u+v=w} \min(\mu_1(u), \mu_2(v))$, and their product $w = u \cdot v$ is described by the membership function $\mu(w) = \max_{u,v: u \cdot v=w} \min(\mu_1(u), \mu_2(v))$.

It is known that these operations can be equivalently described in terms of the α-*cuts*. An α-cut of a fuzzy number $\mu(u)$ is defined as an interval $\mathbf{u}(\alpha) = [u^-(\alpha), u^+(\alpha)]$, where

$$u^-(\alpha) \overset{\text{def}}{=} \inf\{u : \mu(u) \geq \alpha\} \text{ and } u^+(\alpha) \overset{\text{def}}{=} \sup\{u : \mu(u) \geq \alpha\}. \qquad (6.1.26)$$

The α-cuts corresponding to the sum $w = u + v$ can be described, for every α, as

$$[w^-(\alpha), w^+(\alpha)] = [u^-(\alpha), u^+(\alpha)] + [v^-(\alpha), v^+(\alpha)], \qquad (6.1.27)$$

or, equivalently, as

$$[w^-(\alpha), w^+(\alpha)] = [u^-(\alpha) + v^-(\alpha), u^+(\alpha) + v^+(\alpha)]. \qquad (6.1.28)$$

Similarly, the α-cuts corresponding to the product $w = u \cdot v$ can be described as

$$[w^-(\alpha), w^+(\alpha)] = [u^-(\alpha), u^+(\alpha)] \cdot [v^-(\alpha), v^+(\alpha)]. \qquad (6.1.29)$$

If both fuzzy numbers u and v are non-negative (e.g., if they are limited to the interval $[0, 1]$), then the α-cuts corresponding to the product can be described as

$$[w^-(\alpha), w^+(\alpha)] = [u^-(\alpha) \cdot v^-(\alpha), u^+(\alpha) \cdot v^+(\alpha)]. \qquad (6.1.30)$$

Let us use these definitions to define fair price of fuzzy numbers. Let us start with describing the fair price of fuzzy numbers. Similarly to the interval case, a natural requirement is monotonicity: if for all α, we have $s^-(\alpha) \leq t^-(\alpha)$ and $s^+(\alpha) \leq t^+(\alpha)$, then the fair price of t should be larger than or equal to the fair price of s. It is also reasonable to require continuity: that small changes in $\mu(u)$ should lead to small changes in the fair price.

Definition 6.1.18 By a *fair price under fuzzy uncertainty*, we mean a function $P(s)$ that assigns, to every fuzzy number s, a real number, and which satisfies the following properties:

- if a fuzzy number s is located between $\underline{u}$ and $\overline{u}$, then $\underline{u} \leq P(s) \leq \overline{u}$ (conservativeness);
- if a fuzzy number w is the sum of fuzzy numbers u and v, then we have $P(w) = P(u) + P(v)$ (additivity);
- if for all α, we have

$$s^-(\alpha) \leq t^-(\alpha) \text{ and } s^+(\alpha) \leq t^+(\alpha), \qquad (6.1.31)$$

then we have $P(s) \leq P(t)$ (monotonicity);
- if a sequence of membership functions μ_n uniformly converges to μ, then we should have $P(\mu_n) \rightarrow P(\mu)$ (continuity).

We will see that the fair price of a fuzzy number is described in terms of a Riemann-Stieltjes integral. For readers who need a reminder of what this integral is, a brief reminder is presented at the end of this section.

Proposition 6.1.12 *For a fuzzy number s with a continuous membership function $\mu(x)$, α-cuts $[s^-(\alpha), s^+(\alpha)]$ and a point s_0 at which $\mu(s_0) = 1$, the fair price is equal to*

$$P(s) = s_0 + \int_0^1 k^-(\alpha) \, ds^-(\alpha) - \int_0^1 k^+(\alpha) \, ds^+(\alpha), \tag{6.1.32}$$

for appropriate functions $k^-(\alpha)$ and $k^+(\alpha)$.

Discussion. When the function $g(x)$ is differentiable, the Riemann-Stieltjes integral $\int_a^b f(x) \, dg(x)$ is equal to the usual integral

$$\int_a^b f(x) \cdot g'(x) \, dx,$$

where $g'(x)$ denotes the derivative. When the function $f(x)$ is also differentiable, we can use integration by part and get yet another equivalent form

$$f(b) \cdot g(b) - f(a) \cdot g(a) + \int_a^b F(x) \cdot g(x) \, dx, \tag{6.1.33}$$

with $F(x) = -f'(x)$. In general, a Stieltjes integral can be represented in a similar form for some *generalized function* $F(x)$ (see, e.g., [44]; generalized function are also known as *distributions*; we do not use this term to avoid confusion with probability distributions). Thus, the above general formula can be described as

$$P(s) = \int_0^1 K^-(\alpha) \cdot s^-(\alpha) \, d\alpha + \int_0^1 K^+(\alpha) \cdot s^+(\alpha) \, d\alpha \tag{6.1.34}$$

for appropriate generalized functions $K^-(\alpha)$ and $K^+(\alpha)$.

Conservativeness means that for a crisp number located at s_0, we should have $P(s) = s_0$. For the above formula, this means that

$$\int_0^1 K^-(\alpha) \, d\alpha + \int_0^1 K^+(\alpha) \, d\alpha = 1. \tag{6.1.35}$$

For a fuzzy number which is equal to the interval $[\underline{u}, \overline{u}]$, the above formula leads to

$$P(s) = \left(\int_0^1 K^-(\alpha) \, d\alpha \right) \cdot \underline{u} + \left(\int_0^1 K^+(\alpha) \, d\alpha \right) \cdot \overline{u}. \tag{6.1.36}$$

Thus, Hurwicz optimism-pessimism coefficient α_H is equal to $\int_0^1 K^+(\alpha)\,d\alpha$. In this sense, the above formula is a generalization of Hurwicz's formula to the fuzzy case.

Proof

1°. For every two real numbers $u \geq 0$ and $\gamma \in [0, 1]$, let us define a fuzzy number $s_{\gamma,u}(x)$ with the following membership function: $\mu_{\gamma,u}(0) = 1$, $\mu_{\gamma,u}(x) = \gamma$ for $x \in (0, u]$, and $\mu_{\gamma,u}(x) = 0$ for all other x. For this fuzzy number, α-cuts have the following form: $\mathbf{s}_{\gamma,u}(\alpha) = [0, 0]$ for $\alpha > \gamma$, and $\mathbf{s}_{\gamma,u}(\alpha) = [0, u]$ for $\alpha \leq \gamma$.

Based on the α-cuts, one can easily check that $s_{\gamma,u+v} = s_{\gamma,u} + s_{\gamma,v}$. Thus, due to additivity, $P(s_{\gamma,u+v}) = P(s_{\gamma,u}) + P(s_{\gamma,v})$. Due to monotonicity, the value $P(s_{\gamma,u})$ monotonically depends on u. Thus, similarly to the proof of Proposition 6.1.1, we can conclude that $P(s_{\gamma,u}) = k^+(\gamma) \cdot u$ for some value $k^+(\gamma)$.

By definition, the fuzzy number $s_{\gamma,u}$ is located between 0 and u, so, due to conservativeness, we have $0 \leq P(s_{\gamma,u}) \leq u$ for all u. This implies that $0 \leq k^+(\gamma) \leq 1$.

2°. Let us now consider a fuzzy number s whose membership function is equal to 0 for $x < 0$, jumps to 1 for $x = 0$, and then continuously decreases to 0. For this fuzzy number, all α-cuts have the form $[0, s^+(\alpha)]$ for some $s^+(\alpha)$. By definition of an α-cut, the value $s^+(\alpha)$ decreases with α.

For each sequence of values

$$\alpha_0 = 0 < \alpha_1 < \alpha_2 < \cdots < \alpha_{n-1} < \alpha_n = 1,$$

we can define a fuzzy number s_n with the following α-cuts $[s_n^-(\alpha), s_n^+(\alpha)]$:

- $s_n^-(\alpha) = 0$ for all α; and
- when $\alpha \in [\alpha_i, \alpha_{i+1})$, then $s_n^+(\alpha) = s^+(\alpha_i)$.

Since the membership function of s is continuous, when $\max(\alpha_{i+1} - \alpha_i) \to 0$, we have $s_n \to s$, and thus, $P(s_n) \to P(s)$.

One can check that the fuzzy number s_n can be represented as a sum of n fuzzy numbers

$$s_n = s_{\alpha_{n-1},s^+(\alpha_{n-1})} + s_{\alpha_{n-2},s^+(\alpha_{n-2})-s^+(\alpha_{n-1})} + \cdots + s_{\alpha_1,\alpha_1-\alpha_2}. \tag{6.1.37}$$

Thus, due to additivity, we have

$$P(s_n) = P(s_{\alpha_{n-1},s^+(\alpha_{n-1})}) + P(s_{\alpha_{n-2},s^+(\alpha_{n-2})-s^+(\alpha_{n-1})}) + \cdots + P(s_{\alpha_1,\alpha_1-\alpha_2}). \tag{6.1.38}$$

Substituting the expression for $P(s_{\gamma,u})$ from Part 1 of this proof, we conclude that

$$P(s_n) = k^+(\alpha_{n-1}) \cdot s^+(\alpha_{n-1}) + k^+(\alpha_{n-2}) \cdot (s^+(\alpha_{n-2}) - s^+(\alpha_{n-1})) + \cdots +$$

$$k^+(\alpha_1) \cdot (\alpha_1 - \alpha_2). \tag{6.1.39}$$

The right-hand side is minus the integral sum for the Riemann-Stieltjes integral $\int_0^1 k^+(\gamma) ds^+(\gamma)$. Since we have $P(s_n) \to P(s)$, this means that the integral sums always converges, the Riemann-Stieltjes integral is defined, and the limit $P(s)$ is equal to this integral.

3°. Similarly, for fuzzy numbers s whose membership function $\mu(x)$ continuously increases from 0 to 1 as x increases to 0 and is equal to 0 for $x > 0$, the α-cuts are equal to $[s^-(\alpha), 0]$, and $P(s) = \int_0^1 k^-(\gamma) ds^-(\gamma)$ for an appropriate function $k^-(\gamma)$.

4°. A general fuzzy number g, with α-cuts $[g^-(\alpha), g^+(\alpha)]$ and a point g_0 at which $\mu(g_0) = 1$, can be represented as the sum of three fuzzy numbers:

- a crisp number g_0;
- a fuzzy number whose α-cuts are equal to

$$[0, g^+(\alpha) - g_0]; \text{ and}$$

- a fuzzy number whose α-cuts are equal to

$$[g_0 - g^-(\alpha), 0].$$

By conservativeness, the fair price of the crisp number is equal to g_0. The fair prices of the second and the third fuzzy numbers can be obtained by using the formulas from Parts 2 and 3 of this proof. By additivity, the fair price of the sum is equal to the sum of the prices. By taking into account that for every constant g_0, $d(g(x) - g_0) = dg(x)$ and thus,

$$\int f(x) d(g(x) - g_0) = \int f(x) dg(x),$$

we get the desired expression.

6.1.12 Case of Z-Number Uncertainty

In this case, we have two fuzzy numbers: the fuzzy number s which describes the values and the fuzzy number p which describes our degree of confidence in the piece of information described by s.

Definition 6.1.19 By a *fair price under Z-number uncertainty*, we mean a function $P(s, p)$ that assigns, to every pair of two fuzzy numbers s and p such that p is located on an interval $[p_0, 1]$ for some $p_0 > 0$, a real number, and which satisfies the following properties:

- if a fuzzy number s is located between $\underline{u}$ and $\overline{u}$, then $\underline{u} \le P(s, 1) \le \overline{u}$ (conservativeness);

- if $w = u + v$ and $r = p \cdot q$, then

$$P(w, r) = P(u, p) + P(v, q) \qquad (6.1.40)$$

(additivity);
- if for all α, we have

$$s^-(\alpha) \leq t^-(\alpha) \text{ and } s^+(\alpha) \leq t^+(\alpha), \qquad (6.1.41)$$

then we have $P(s, 1) \leq P(t, 1)$ (monotonicity);
- if $s_n \to s$ and $p_n \to p$, then $P(s_n, p_n) \to P(s, p)$ (continuity).

Proposition 6.1.13 *For a fuzzy number s with α-cuts $[s^-(\alpha), s^+(\alpha)]$ and a fuzzy number p with α-cuts $[p^-(\alpha), p^+(\alpha)]$, we have*

$$P(s, p) = \int_0^1 K^-(\alpha) \cdot s^-(\alpha) \, d\alpha + \int_0^1 K^+(\alpha) \cdot s^+(\alpha) \, d\alpha +$$

$$\int_0^1 L^-(\alpha) \cdot \ln(p^-(\alpha)) \, d\alpha + \int_0^1 L^+(\alpha) \cdot \ln(p^+(\alpha)) \, d\alpha \qquad (6.1.42)$$

for appropriate generalized functions $K^\pm(\alpha)$ and $L^\pm(\alpha)$.

Proof Due to additivity, we have

$$P(s, p) = P(s, 1) + P(0, p).$$

We already know the expression for $P(s, 1)$; we thus need to find the expression for $P(0, p)$. For logarithms, we have $\ln(p \cdot q) = \ln(p) + \ln(q)$, so in terms of logarithms, additivity takes the usual form

$$P(0, \ln(p) + \ln(q)) = P(0, \ln(p)) + P(0, \ln(q)). \qquad (6.1.43)$$

Thus, similarly to the proof of Proposition 6.1.12, we conclude that

$$P(0, p) = \int_0^1 L^-(\alpha) \cdot \ln(p^-(\alpha)) \, d\alpha + \int_0^1 L^+(\alpha) \cdot \ln(p^+(\alpha)) \, d\alpha. \qquad (6.1.44)$$

By applying additivity to this expression and to the known expression for $P(s, 1)$, we get the desired formula.

Summary. In this section, for different types of uncertainty, we derive the formulas for the fair prices under reasonable conditions of conservativeness, monotonicity, and additivity.

In the simplest case of interval uncertainty, when we only know the interval $[\underline{u}, \overline{u}]$ of possible values of the gain u, the fair price is equal to

$$P([\underline{u}, \overline{u}]) = \alpha_H \cdot \overline{u} + (1 - \alpha_H) \cdot \underline{u},$$

for some parameter $\alpha_H \in [0, 1]$. Thus, the fair price approach provides a justification for the formula originally proposed by a Nobelist L. Hurwicz, in which α_H describes the decision maker's optimism degree: $\alpha_H = 1$ corresponds to pure optimism, $\alpha_H = 0$ to pure pessimism, and intermediate values of α_H correspond to a realistic approach that takes into account both best-case (optimistic) and worst-case (pessimistic) scenarios.

In a more general situation, when the set S of possible values of the gain u is not necessarily an interval, the fair price is equal to

$$P(S) = \alpha_H \cdot \sup S + (1 - \alpha_H) \cdot \inf(S).$$

If, in addition to the set S of possible values of the gain u, we also know a subset $s \subseteq S$ of "most probable" gain values, then the fair price takes the form

$$P(S, s) = \inf s + \alpha_u \cdot (\sup s - \inf s) + \alpha_U \cdot (\sup S - \sup s) + \alpha_L \cdot (\inf S - \inf s),$$

for some values α_u, α_L, and α_U from the interval $[0, 1]$. In particular, when both sets S and s are intervals, i.e., when $S = [\underline{u}, \overline{u}]$ and $s = [\underline{m}, \overline{m}]$, the fair price takes the form

$$P([\underline{u}, \overline{u}], [\underline{m}, \overline{m}]) = \underline{m} + \alpha_u \cdot (\overline{m} - \underline{m}) + \alpha_U \cdot (\overline{u} - \overline{m}) + \alpha_L \cdot (\underline{u} - \underline{m}).$$

When the interval s consists of a single value u_0, this formula turns into

$$P([\underline{u}, \overline{u}], u_0) = \alpha_L \cdot \underline{u} + (1 - \alpha_L - \alpha_U) \cdot u_0 + \alpha_U \cdot \overline{u}.$$

When, in addition to the set S, we also know the cumulative distributive function (cdf) $F(u)$ that describes the probability distribution of different possible values u, then the fair price is equal to the expected value of the gain

$$P(F) = \int u \, dF(u).$$

In situations when for each u, we only know the interval $[\underline{F}(u), \overline{F}(u)]$ of possible values of the cdf $F(u)$, then the fair price is equal to

$$P([\underline{F}, \overline{F}]) = \alpha_H \cdot \int u \, d\overline{F}(u) + (1 - \alpha_H) \cdot \int u \, d\underline{F}(u).$$

Finally, when uncertainty is described by an improper interval $[\underline{u}, \overline{u}]$ with $\underline{u} > \overline{u}$, the fair price is equal to

$$P([\underline{u}, \overline{u}]) = \alpha_H \cdot \overline{u} + (1 - \alpha_H) \cdot \underline{u}.$$

Similar formulas are provided for the case of fuzzy uncertainty.

Riemann-Stieltjes integral—reminder. As promised, let us recall what is the Riemann-Stieltjes integral. This integral is a natural generalization of the usual (Riemann) integral.

In general, an intuitive meaning of a Riemann integral $\int_a^b f(x)\,dx$ is that it is an area under the curve $y = f(x)$. To compute this integral, we select points

$$a = x_1 < x_2 < \cdots < x_{n-1} < x_n = b,$$

and approximate the curve by a piece-wise constant function $\widetilde{f}(x) = f(x_i)$ for $x \in [x_i, x_{i+1})$. The subgraph of this piece-wise constant function is a union of several rectangles, so its area is equal to the sum of the areas of these rectangles $\sum f(x_i) \cdot (x_{i+1} - x_i)$. This sum is known as the *integral sum* for the integral $\int_a^b f(x)\,dx$. Riemann's integral can be formally defined as a limit of such integral sums when $\max(x_{i+1} - x_i) \to 0$.

A Riemann-Stieltjes integral $\int_a^b f(x)\,dg(x)$ is similarly defined as the limit of the sums $\sum f(x_i) \cdot (g(x_{i+1}) - g(x_i))$ when $\max(x_{i+1} - x_i) \to 0$.

6.2 Decision Making Under Uncertainty and Restrictions on Computation Resources: Educational Case Studies

In the previous section, we described the general ideas of optimal decision making under uncertainty. In this section, we illustrate these ideas on three examples of decisions related to education. In the first subsection, we analyze how to assign a grade for a partially solved problem. In the second subsection, we analyze how to to take into account a student's degree of confidence when evaluating test results. In the final third subsection, we analyze how to combine grades for different classes into a single number measuring the student's knowledge.

6.2.1 How to Assign a Grade for a Partially Solved Problem: Decision-Making Approach to Partial Credit

When a student performed only some of the steps needed to solve a problem, this student gets partial credit. This partial credit is usually proportional to the number of stages that the student performed. This may sound reasonable, but in engineering education, this leads to undesired consequences: for example, a student who did not solve any of the 10 problems on the test, but who successfully performed 9 out of 10 stages needed to solve each problem will still get the grade of A ("excellent"). This may be a good evaluation of the student's intellectual ability, but for a engineering company that hires this A-level student, this will be an unexpected disaster. In this subsection, we analyze this problem from the viewpoint of potential loss to a company, and we show how to assign partial credit based on such loss estimates. Our conclusion is that this loss (and thus, the resulting grade) depend on the size of the engineering company. Thus, to better understand the student's strengths, it is desirable, instead of a single overall grade, to describe several grades corresponding to different company sizes.

Results from this subsection first appeared in [93]

Need to assign partial credit.

- If on a test, a problem is solved correctly, then the student gets full credit.
- If the student did not solve this problem at all, or the proposed solution is all wrong, the student gets no credit for this problem.

In many cases, the student correctly performed some steps that leads to the solution, but, due to missing steps, still did not get the solution. To distinguish these cases from the cases when the student did not perform any steps at all, the student is usually given partial credit for this problem.

The more steps the student performed, the more partial credit this student gets. From the pedagogical viewpoint, partial credit is very important: it enables the student who is learning—but who has not yet reached a perfect-knowledge stage—to see his or her progress; see, e.g., [10, 54].

How partial credit is usually assigned. Usually, partial credit is assigned in a very straightforward way:

- if the solution to a problem requires n steps,
- and $k < n$ of these steps have been correctly performed,
- then we assign the fraction $\dfrac{k}{n}$ of the full grade.

   For example:

- if a 10-step problem is worth 100 points and
- a student performed 9 steps correctly,
- then this student gets 90 points for this problem.

For engineering education, this usual practice is sometimes a problem.

- If our objective is simply to check intellectual progress of a student, then the usual practice of assigning partial credit makes perfect sense,
- However, in engineering education, especially in the final classes of this education, the goal is to check how well a student is prepared to take on real engineering tasks.

Let us show that from this viewpoint, the usual practice is not always adequate.

Let us consider a realistic situation when, out of 10 problems on the test, none of these problems were solved by the student. However, if

- in each of these ten problems,
- the student performed 9 stages out of 10,
- this student will get 9 points of the ten for this problem.

Thus, the overall number of points for the test is $10 \cdot 9 = 90$, and the student gets the grade of A ("excellent") for this test.

None of the original ten problem are solved, but still the student got an A. This does not sound right, especially when we compare it with a different student, who:

- correctly solved 9 problems out of 10, but
- did not even start solving the tenth problem.

This second student will also get the same overall grade of 90 out of 100 for this test. However, if we assume that this test simulates the real engineering situation, the second student clearly performed much better.

This example shows that for engineering education, we need to come up with a different scheme of assigning partial credit.

What we do in this subsection. In this subsection, we analyze the problem, and come up with an appropriate scheme for assigning partial credit.

Description of the situation. To appropriately analyze the problem, let us imagine that this student:

- has graduated and
- is working for an engineering company.

In such a situation, a natural way to gauge the student's skill level is to estimate the overall benefit that he or she will bring to this company.

Let us start with considering a single problem whose solution consists of n stages. Let us assume that the newly hired student can correctly perform k out of these n stages. This is not a test, this is a real engineering problem, it needs to be solved, so someone else must help to solve the remaining $n - k$ stages.

Possible scenarios. From the viewpoint of benefit to a company, it is reasonable to distinguish between two possible situations:

- It is possible that in this company, there are other specialists who can help with performing the remaining $n - k$ stages. In this case, the internal cost of this additional (unplanned) help is proportional to the number of stages.
- It is also possible that no such specialists can be found within a company (this is quite probable if we have a small company). In this case, we need to hire outside help. In such a situation, the main part of the cost is usually the hiring itself: the consultant needs to be brought in, and

  – the cost of bringing in the consultant
  – is usually much higher than the specific cost of the consultant performing the corresponding tasks.

Estimating expected loss to a company: case of a single problem. We want to gauge the student's grade based on the financial implication of his or her imperfect knowledge on the company that hires this student.

We cannot exactly predict these implications, since we do not know for sure whether the company will have other specialists in-house who can fill in the stages that the newly hired student is unable to perform. Thus, we have a situation of uncertainty. According to decision theory (see, e.g., [35, 97, 133]), in such situations, a reasonable decision maker should select an alternative with the largest value of expected utility. Thus, a reasonable way to gauge the student's effect on the company is to estimate the expected loss caused by the student's inability to perform some tasks.

To estimate this expected loss, we need to estimate:

- the losses corresponding to both above scenarios, and
- the probability of each of these scenarios.

As we have mentioned, in the first scenario, when all the stages are performed in-house, the cost is proportional to the number of stages. So, if we denote the cost of performing one stage in-house by c_i (i for "in-house"), the resulting cost is equal to $c_i \cdot (n - k)$.

In the second scenario, when we have to seek a consultant's help, as we also mentioned, the cost practically does not depend on the number of stages. Let us denote this cost by c_h (h for "help").

To complete the estimate, we need to know the probabilities of the two scenarios. Let p denote the probability that inside the company, there is a specialist that can help with performing a given stage. It is reasonable to assume that different stages are independent from this viewpoint, so the overall probability that we can find inside help for all $n - k$ stages is equal to p^{n-k}. Thus:

- with probability p^{n-k}, we face the first (in-house) scenario, with the cost $c_i \cdot (n - k)$;
- with the remaining probability $1 - p^{n-k}$, we face the second (outside help) scenario, with the cost c_h.

The resulting expected loss is thus equal to

$$c_i \cdot p^{n-k} \cdot (n - k) + c_h \cdot \left(1 - p^{n-k}\right). \qquad (6.2.1)$$

Estimating expected loss to a company: case of several problems. Several problems on a test usually simulate different engineering situations that may occur in real life. We can use the formula (6.2.1) to estimate the loss caused by each of the problems.

Namely, let n be the average number of stages of each problem. Then, if in the j-th problem, the student can successfully solve k_j out of n problems, then the expected loss is equal to:

$$c_i \cdot p^{n-k_j} \cdot \left(n - k_j\right) + c_h \cdot \left(1 - p^{n-k_j}\right). \qquad (6.2.2)$$

The overall expected loss L can be then computed as the sum of the costs corresponding to all J problems, i.e., as the sum

$$L = \sum_{j=1}^{J} \left(c_i \cdot p^{n-k_j} \cdot \left(n - k_j \right) + c_h \cdot \left(1 - p^{n-k_j} \right) \right). \qquad (6.2.3)$$

So how should we assign partial credit. Usually, the credit is counted in such as way that complete knowledge corresponds to 100 points, and a complete lack of knowledge corresponds to 0 points. In this case, to assign partial credit, we should subtract, from the ideal case of 100 point, an amount proportional to the expected loss caused by the student's lack of skills.

In other words, the grade g assigned to the student should be equal to

$$g = 100 - c \cdot L, \qquad (6.2.4)$$

for an appropriate coefficient of proportionality c. The corresponding c should be selected in such a way that for a complete absence of knowledge, we should subtract exactly 100 points.

The complete lack of knowledge corresponds to the case when for each problem j, the student is not able to solve any stage, i.e., when $k_j = 0$ for all j. In this case, the formula (6.2.3) takes the form

$$\overline{L} = J \cdot (c_i \cdot p^n \cdot n + c_h \cdot (1 - p^n)). \qquad (6.2.5)$$

We want to select the coefficient of proportionality c in such a way that this worsecase will be equal to 100: $c \cdot \overline{L} = 100$. From this equality, we conclude that

$$c = \frac{100}{\overline{L}} = \frac{100}{J \cdot (c_i \cdot p^n \cdot n + c_h \cdot (1 - p^n))}. \qquad (6.2.6)$$

Substituting this expression and the expression (6.2.3) for L into the formula (6.2.4), we conclude that

$$g = 100 - 100 \cdot \frac{\sum_{j=1}^{J} \left(c_i \cdot p^{n-k_j} \cdot \left(n - k_j \right) + c_h \cdot \left(1 - p^{n-k_j} \right) \right)}{J \cdot (c_i \cdot p^n \cdot n + c_h \cdot (1 - p^n))}. \qquad (6.2.7)$$

We can simplify this expression if we divide both numerator and denominator of this fraction by the factor c_h. In this case, this factor c_h disappears in terms proportional to c_h, and terms proportional to c_i become now proportional to the ratio

$$c_i' \stackrel{\text{def}}{=} \frac{c_i}{c_h} \ll 1.$$

As a result, we arrive at the following formula.

How to assign partial credit: the resulting formula. Our analysis shows that for a student who, for each j-th problem out of J, performed k_j out of n stages, should be given the following grade:

$$g = 100 - 100 \cdot \frac{\sum_{j=1}^{J} \left(c_i' \cdot p^{n-k_j} \cdot (n - k_j) + (1 - p^{n-k_j})\right)}{J \cdot (c_i' \cdot p^n \cdot n + (1 - p^n))}. \qquad (6.2.8)$$

Here c_i' is the ratio of an in-house cost of performing a stage to the cost of hiring an outside consultant.

Different types of companies. The above formula use a parameter: namely, the probability p that it is possible to perform a stage in-house, without hiring outside help. We have already mentioned that the value of this parameter depends on the company size:

- In a very big company, with many engineers of different type, this probability is close to 1.
- On the other hand, in a small company this probability is very small.

On these two extreme cases, let us illustrate the use our formula (6.2.8).

First extreme case: a very big company. In this case, when $p = 1$, we have $1 - p^{n-k_j} = 1 - p^n = 0$, so the formula (6.2.8) takes the simplified form

$$g = 100 - 100 \cdot \frac{\sum_{j=1}^{J} c_i' \cdot (n - k_j)}{J \cdot c_i' \cdot n}. \qquad (6.2.9)$$

Dividing both numerator and denominator by $c_i' \cdot n$, we get

$$g = 100 - 100 \cdot \frac{\sum_{j=1}^{J} \left(1 - \frac{k_j}{n}\right)}{J}, \qquad (6.2.10)$$

i.e., the formula

$$g = 100 \cdot \frac{1}{J} \cdot \sum_{j=1}^{J} \frac{k_j}{n}. \qquad (6.2.11)$$

This is the usual formula for assigning partial credit—thus, this formula corresponds to the case when a student is hired by a very big company.

Second extreme case: a very small company. In this case, $p = 0$, i.e., every time a student is unable to solve the problem, the company has to hire an outside help. The cost of outside help does not depend on how many stages the student succeeded in solving: whether a student performed all but one stages or none, the loss is the same.

In this case, the student gets no partial credit at all—if the answer is not correct, there are 0 points assigned to this student.

Comment. This situation is similar to how grades are estimated on the final exams in medicine: there, an "almost correct" (but wrong) answer can kill the patient, so such "almost correct" answers are not valued at all.

Intermediate case: general description. In the intermediate cases, we do assign some partial credit, but this credit is much smaller than in the traditional assignment (which, as we have shown, corresponds to the large-company case).

Intermediate case: towards an approximate formula. We have already mentioned that $c_i \ll c_h$ and thus, $c_i' \ll 1$. Thus, we can safely ignore the terms proportional to c_i' in our formula (6.2.8). As a result, we get the following approximate formula:

$$g = 100 \cdot \left(1 - \frac{\sum\limits_{j=1}^{J} \left(1 - p^{n-k_j} \right)}{J \cdot (1 - p^n)} \right).$$

If we subtract the fraction from 1, we get

$$g = 100 \cdot \frac{J \cdot (1 - p^n) - \sum\limits_{j=1}^{J} \left(1 - p^{n-k_j} \right)}{J \cdot (1 - p^n)},$$

i.e.,

$$g = 100 \cdot \frac{1}{J} \cdot \sum_{j=1}^{J} \frac{p^{n-k_j} - p_n}{1 - p^n}. \tag{6.2.12}$$

In other words, the overall grade for the test is equal to the average of the grades g_j for all the problems:

$$g = 100 \cdot \frac{1}{J} \cdot \sum_{j=1}^{J} g_j, \tag{6.2.13}$$

where

$$g_j \overset{\text{def}}{=} \frac{p^{n-k_j} - p_n}{1 - p^n}. \tag{6.2.14}$$

Usually, the number of stages p is reasonably large, so $p^n \approx 0$, and we arrive at the following formulas.

Intermediate case: resulting approximate formulas. The grade for the test can be described as an average of the grades g_j for individual problems, where for each problem, if a student performed k_j steps out of n, the grade is:

$$g_j \approx p^{n-k_j}. \tag{6.2.15}$$

Comment. A similar formula can be obtained if we consider a more general case, when different problems j may have different stages n_j. In this case, the general formula has the form

$$g = 100 - 100 \cdot \frac{\sum\limits_{j=1}^{J} \left(c_i \cdot p^{n_j - k_j} \cdot \left(n_j - k_j \right) + c_h \cdot \left(1 - p^{n_j - k_j} \right) \right)}{\sum\limits_{j=1}^{J} \left(c_i \cdot p^{n_j} \cdot n_j + c_h \cdot \left(1 - p^{n_j} \right) \right)}, \tag{6.2.16}$$

and the corresponding approximate formula reduced to the average of the grades

$$g_j \approx p^{n_j - k_j}. \tag{6.2.17}$$

So what do we propose. Our analysis shows that, depending on the size of company, we should assign partial credit differently. So maybe this is a way to go:

- instead of trying to describe the student's knowledge by a single number,
- we use different numbers corresponding to several different values of the parameter p (i.e., in effect, depending on the size of the company).

For example, we can select values $p = 0$, $p = 0.1$, ..., $p = 0.9$, and $p = 1$. Or, as a start, we can select just values $p = 0$ and $p = 1$ (maybe supplemented with $p = 0.5$).

   We recommend that all these grades should be listed in the transcript, and each company should look into the grade that is the best fit for their perceived value p.

Comment. In this subsection, we assumed that the student is absolutely confident about his or her answers. In real life, the student is often not fully confident. The corresponding degree of confidence should also be taken into account when gauging the student's knowledge—i.e., when assigning partial credit; possible ways to take this uncertainty into account are described in the next subsection.

Need for an intuitive interpretation of the above formula. Our formula for partial credit comes from a simplified—but still rather complicated—mathematical model. Since the model is simplified, we cannot be 100% sure that a more complex model would not lead to a different formula. To make this formula more acceptable, we need to supplement the mathematical derivation with a more intuitive explanation. Such an explanation will be provided in this section and in the following one; specifically:

- in this section, we will show that the above formula is in good accordance with the general well-accepted ideas of computational intelligence;
- in the following section, we show that this formula is also in good accordance with a common sense analysis of the situation.

Let us analyze the problem of partial credit from the viewpoint of computational intelligence. The grade for a problem can be viewed as the instructor's degree of confidence that the student knows the material. The only information that we can use to compute this degree is that:

- the student successfully performed k_j stages, and
- the student did not perform the remaining $n - k_j$ stages.

Let us start our analysis with the simplest case, when the problem has only one stage. In this case:

- if the student successfully performed this stage, then our degree of confidence in this student's knowledge is higher, while
- if the student did not perform this stage, our degree of confidence in this student's knowledge is lower.

Let D denote our degree of confidence in the student's knowledge when the student successfully performed the stage, and let d ($d < D$) be the degree of confidence corresponding to the case when the student did not succeed in performing this stage.

In general, each of the k_j successful stages add a confidence D, and each of $n - k_j$ unsuccessful stages add a confidence degree d. The need for combining different degrees of confidence is well-studied in fuzzy logic (see, e.g., [65, 120, 163]), where the degree of confidence in an "and"-statement A and B is estimated by applying an appropriate "and"-operation (a.k.a. t-norm) to the degrees of confidence in individual statements.

The two simplest (and most widely used) "and"-operations are min and product. Let us consider what will happen if we use each of these operations.

If we use the minimum "and"-operation, there will be no partial credit. Let us first analyze what will happen if we use min as the "and"-operation. In this case, we have only two possible results of applying this "and"-operation:

- if the student successfully performed all the stages of solving a problem, then the resulting degree of confidence is equal to

$$\min(D, \ldots, D) = D;$$

- on the other hand, if the student did not succeed in at least one stage, then, no matter how many stages were missed, we get

$$\min(D, \ldots, D, d, \ldots, d) = d.$$

So, in this case, we:

- either give the student the full credit—if the student has successfully performed all the stages,
- or, if the student failed in at least one stage of the problem, we give the student the exact same grade as if this student failed in all the stages.

Thus, if we use the minimum "and"-operation, we do not assign any partial credits—and, as we have mentioned earlier, partial credit is pedagogically important.

What if we use the product "and"-operation. If we use the product "and"-operation, then the resulting degree of confidence is equal to

$$\mu = D \cdot \cdots \cdot D \ (k_j \text{ times}) \cdot d \cdot \cdots \cdot d \ (n - k_j \text{ times}) = D_j^k \cdot d^{n-k_j}. \quad (6.2.18)$$

As usual in fuzzy techniques, it is convenient to *normalize* these value, i.e., to divide all these degrees μ by the largest possible degree $\overline{\mu}$—so that after the division, the largest of the new degrees $\mu' = \dfrac{\mu}{\overline{\mu}}$ will be equal to 1.

The largest degree is clearly attained when the student has successfully performed all n stages of solving the problem, i.e., when $k_j = n$ and, correspondingly, $n - k_j = 0$. In this case, formula (6.2.18) leads to $\overline{\mu} = D^n$. Dividing μ by $\overline{\mu}$, we conclude that

$$\mu' = \frac{\mu}{\overline{\mu}} = \frac{D_j^k \cdot d^{n-k_j}}{D^n} = \frac{d^{n-k_j}}{D^{n-k_j}} = p^{n-k_j}, \quad (6.2.19)$$

where we have denoted $p \overset{\text{def}}{=} \dfrac{d}{D}$.

Conclusion: we get the exact same formula for partial credit. We can see that:

- the formula (6.2.19) that we obtained based on the ideas from computational intelligence, and
- the formula (6.2.17) that is obtained based on our mathematical model

are identical. Thus, our formula (6.2.17) is indeed in good accordance with the computational intelligence ideas.

Towards a commonsense interpretation of our partial credit formula. The above justification is based on the general description of different "and"-operations. These general descriptions have nothing to do with the specific problem of assigning partial credit. So maybe for this specific problem we should have used a different "and"-operation, which may lead us to different formula?

To make our formula for partial credit move convincing, it is therefore desirable to supplement that general justification with a commonsense interpretation which would be directly related to the problem of assigning partial credit.

Main idea behind about commonsense justification. Knowledge about each subject can be viewed as an ever-growing tree:

- We start with some very basic facts. We can say that these facts form the basic (first) level of the knowledge tree.
- Then, based on these very basic facts, we develop some concepts of the second level. For example, in Introduction to Computer Science, once the students understood the main idea of an algorithm, they usually start describing different types of variables: integers, real numbers, characters, strings, etc.

- Based on the knowledge of the second level, we then further branch out into different concepts of the third level, etc.

In the first approximation, we can assume that the branching b is the same on each level, so that:

- we have one cluster of concepts on Level 1,
- we have b different concepts (or clusters of concepts) on Level 2,
- we have b^2 different concepts or clusters of concepts on Level 3,
- …
- we have b^{k-1} different concepts or clusters of concepts on each Level k,
- …, and
- we have b^{n-1} different concepts or clusters of concepts on the highest level n.

The overall number of concepts of all levels from 1 to n is equal to the sum

$$1 + b + \cdots + b^{n-1} = \frac{b^n - 1}{b - 1}.$$

For each problem, usually, each of n stages corresponds to the corresponding level:

- the first stage corresponds to Level 1,
- the second stage corresponds to Level 2, …
- the n-th stage corresponds to Level n.

From this viewpoint, when a student was able to only successfully perform k stages, this is an indication that this student have mastered only the concepts of the first k levels.

With this limited knowledge, out of all possible $\dfrac{b^n - 1}{b - 1}$ concepts, the student has mastered only

$$1 + b + \cdots + b^{k-1} = \frac{b^k - 1}{b - 1}$$

of them.

A reasonable way to gauge the student's knowledge is by estimating what is the portion of concepts that the student learned, i.e., by computing the ratio

$$\frac{\dfrac{b^k - 1}{b - 1}}{\dfrac{b^n - 1}{b - 1}} = \frac{b^k - 1}{b^n - 1}. \tag{6.2.20}$$

Here, n is reasonably large and k is also large—if k is the small, this means that the student is failing this class anyway, so it does not matter how exactly we assign partial credit. In this case, $b^n \gg 1$ and $b^k \gg 1$. So, by ignoring 1s in comparison with b^n and b^k, we can come up with the following approximate formula

$$\frac{b^k}{b^n} = \frac{1}{b^{n-k}} = p^{n-k}, \tag{6.2.21}$$

where we have denoted $p \stackrel{\text{def}}{=} \dfrac{1}{b}$.

Summary: we get the exact same formula for partial credit. We can see that:

- the formula (6.2.21) that we obtained based on our commonsense analysis if the partial credit problem, and
- the formula (6.2.17) that is obtained based on our mathematical model

are identical. Thus, our formula (6.2.17) is indeed in good accordance with common sense.

6.2.2   How to Take into Account a Student's Degree of Certainty When Evaluating Test Results

To more adequately gauge the student's knowledge, it is desirable to take into account not only whether the student's answers on the test are correct or nor, but also how confident the students are in their answers. For example, a situation when a student gives a wrong answer, but understands his/her lack of knowledge on this topic, is not as harmful as the situation when the student is absolutely confident in his/her wrong answer. In this subsection, we use the general decision making theory to describe the best way to take into account the student's degree of certainty when evaluating the test results.

Results from this subsection first appeared in [78].

Need to take into account the student's degree of certainty. On a usual test, a student provides answers to several questions, and the resulting grade depends on whether these answers are correct.

However, this approach does not take into account how confident the students is in his/her answer.

In real life situations, when a person needs to make a decision, it may happen that a person does not know the correct answer—but it is not so bad if this person realizes that his knowledge is weak on this subject. In this case, he/she may consult an expert and come up with a correct decision. The situation is much worse if a decision maker is absolutely confident in his/her wrong decision.

For example, we do not expect our medical doctor to be perfect, but what we do expect is that when the doctor is not sure, he/she knows that his/her knowledge of this particular medical situation is lacking, and either consults a specialist him/herself or advises the patient to consult a specialist.

From this viewpoint, when gauging the student's level of knowledge, it is desirable:

- to explicitly ask the student how confident he/she is in the corresponding answer, and
- to take this degree of confidence into account when evaluating the resulting grade.

Some tests already solicit these confidence degrees from the students; see, e.g., [96, 105, 121, 122]; see also [58].

How can we take the student's degree of certainty into account? Once we have the student's degrees of confidence in different answers, how should we combine these degrees of confidence into a single overall grade?

As of now, the existing combination rules are semi-heuristic. It is therefore desirable to come up with well-justified rules for such combination.

What we do in this subsection. In this subsection, we propose to use decision making theory to come up with a combination rule that adequately describes the effect of possible uncertainty and/or wrong answers on the possible decisions.

For that, we emulate a real-life decision making situation, when the decision is made by a group of specialists including the current student. In this setting, we estimate the expected contribution of this student's knowledge to the quality of the resulting decision.

Need to describe how decisions are made: reminder. A natural idea to gauge student's uncertain knowledge is to analyze how this uncertainty affects the decisions. To be able to perform this analysis, we need to describe what is a reasonable way to make a decision based on the opinion of several uncertain (and independent) experts.

Decision making under uncertainty: general case. According to decision theory, decisions made by a rational agent can be described as follows:

- to each possible situation, we assign a numerical value called its *utility*, and
- we select an action for which the expected value of utility is the largest possible.

Let us start with the simplest simplified case. Let us start our analysis with the simplest case, in which:

- we only have one question and
- for this question, there are only two possible alternatives A_1 and A_2.

Let P_1 be the student's degree of confidence in the answer A_1. Since we assumed that there are only two possible answers, the student's degree of confidence in the other answer A_2 is equal to $P_2 = 1 - P_1$.

To gauge the effect of the student's answer on the resulting decision, let us assume that for each of the two alternatives A_1 and A_2, we know the optimal action.

For example, we have two possible medical diagnoses, and for each of these diagnoses, we know an optimal treatment.

Let $u_{i,j}$ be the utility corresponding to the case when

- the actual situation is A_i and

- we use the action which is optimal for the alternative A_j.

In these terms, the fact that the action corresponding to A_1 is optimal for the situation A_1 means that $u_{1,1} > u_{1,2}$; similarly, we get $u_{2,2} > u_{2,1}$.

If we know the probabilities p_1 and $p_2 = 1 - p_1$ of both situations, then we select the action corresponding to A_1 if its expected utility is larger, i.e., if

$$p_1 \cdot u_{1,1} + (1 - p_1) \cdot u_{2,1} \geq p_1 \cdot u_{1,2} + (1 - p_1) \cdot u_{2,2},$$

i.e., equivalently, if

$$p_1 \geq t \stackrel{\text{def}}{=} \frac{u_{2,2} - u_{2,1}}{(u_{1,1} - u_{1,2}) + (u_{2,2} - u_{2,1})}. \tag{6.2.22}$$

If the actual situation is A_1, then the optimal action is the one corresponding to A_1. Thus, the above inequality describes when the optimal action will be applied—when our degree of confidence in A_1 exceeds the above-described threshold t.

How to estimate the probabilities of different alternatives under expert uncertainty. Let us denote the number of experts by n. Let us assume that for each expert k, we know this expert's degree of confidence (subjective probability) $p_{1,k}$ in alternative A_1, and his/her degree of confidence $p_{2,k} = 1 - p_{1,k}$ in alternative A_2.

In general, we do not have prior reasons to believe that some experts are more knowledgeable than others, so we assume that all n experts are equally probable to be right:

$$P(k\text{-th expert is right}) = \frac{1}{n}.$$

Thus, by the law of complete probability, we have

$$p_1 = \text{Prob}(A_1 \text{ is the actual alternative}) =$$

$$\sum_{k=1}^{n} P(k\text{-th expert is right}) \cdot P(A_1 \mid k\text{-th expert is right}),$$

hence

$$p_1 = \frac{1}{n} \cdot \sum_{k=1}^{n} p_{1,k}. \tag{6.2.23}$$

How to estimate the student's contribution to the correct decision. We started with the average of the probabilities of n experts. Once we add the student as a new expert, with probabilities $p_{1,n+1} = P_1$ and $p_{2,n+1} = P_2$, the probability p_1 changes to the new value

$$p_1' = \frac{1}{n+1} \cdot \sum_{k=1}^{n+1} p_{1,k} = \frac{n}{n+1} \cdot p_1 + \frac{1}{n+1} \cdot P_1. \qquad (6.2.24)$$

For large n, this addition is small, so in most cases, it does not change the decision. However:

- sometimes, the increase in the estimated probability p_1 will help us switch from the wrong decision to the correct one, and
- sometimes, vice versa, the new estimate will be smaller than the original one and thus, because of the addition of the student's opinion, the group will switch from the correct to the wrong decision.

According to the general decision theory ideas, the student's contribution can be gauged as the expected utility cased by the corresponding change, i.e., as the probability of the positive change times the gain minus the probability of the negative loss times the corresponding loss.

The probability of a gain is equal to the probability that $p_1 < t$ but $p_1' \geq t$. Due to (6.2.24), the inequality $p_1' \geq t$ is equivalent to

$$p_1 \geq t + \frac{1}{n} \cdot t - \frac{1}{n} \cdot P_1.$$

Thus, the probability of the gain is equal to the probability that the previous estimate p_1 is in the interval

$$\left[t + \frac{1}{n} \cdot t - \frac{1}{n} \cdot P_1, \; t \right].$$

For large n, this interval is narrow, so this probability can be estimated as the probability density $\rho(t)$ of the probability corresponding to p_1 times the width

$$\frac{1}{n} \cdot P_1 - \frac{1}{n} \cdot t$$

of this interval. Thus, this probability is a linear function of P_1.

Similarly, the probability of the loss is also a linear function of P_1 and hence, the expected utility also linearly depends on P_1.

How to gauge student's knowledge: analysis of the problem. The appropriate measure should be a linear function of the student's degree of certainty P_1. So, if we originally assign:

- N points to a student who is absolutely confident in the correct answer and
- 0 points to a student who is absolutely confident in the wrong answer,

then the number of points assigned in general should be a linear function of P_1 that is:

- equal to N when $P_1 = 1$, and

- equal to 0 when $P_1 = 0$.

One can check that the only linear function with this property is the function $N \cdot P_1$. Thus, we arrive at the following recommendation:

How to gauge student's knowledge: the resulting recommendation. When a student supplies his/her degree of confidence P_1 in the answer A_1 (and, correspondingly, the degree of confidence $P_2 = 1 - P_1$ in the answer A_2), then we should give the student:

- $N \cdot P_1$ points if A_1 is the correct answer, and
- $N \cdot P_2$ points if A_2 is the correct answer,

where N is the number of points that the student would get for an absolutely correct answer with confidence 1.

Discussion. Let us show that if we follow the above recommendation, then we assign different numbers of grades in two situations that we wanted to distinguish:

- the bad situation in which a student is absolutely confident in the wrong answer $p_1 = 0$ and $p_2 = 1$, and
- a not-so-bad situation when a student is ignorant but understands his or her ignorance and assigns the degree of confidences $p_1 = p_2 = 0.5$ to both possible answers.

Indeed:

- In the first (bad) situation, the student gets the smallest number of points: 0.
- In the second, not-so-bad situation, the student gets $N \cdot 0.5$ points, which is more than 0.

Comment. Of course, if we assign the points this way, the fact that someone with no knowledge can get 50% means that we need to appropriately change the thresholds for A, B, and C grades.

What if a question has several possible answers: analysis of the problem. In general, a question can have several possible answers corresponding to several alternatives. Let us denote these alternatives by $A_1, \ldots, A_s$.

In this case, a student assigns, to each of these alternatives A_i, his/her degree of certainty P_i. Since we know that exactly one of the given s alternatives is true, these probabilities should add up to 1: $\sum_{i=1}^{s} P_i = 1$.

How does adding this student's knowledge change the decision of n experts? Similarly to the previous case:

- it may be that previously, the experts selected a wrong alternative, and the student's knowledge can help select the correct alternative A_1;
- it also may be that the experts selected the correct alternative, but the addition of the student's statement will lead to the selection of a wrong alternative.

Let us describe this in detail.

Similarly to the case of two alternative, we can conclude that a group of experts selects an action corresponding to the alternative A_{i_0} if the corresponding expected utility is larger than the expected utility of selecting any other action, i.e., if

$$\left(\sum_{j=1}^{s} p_j \cdot u_{i_0,j} \right) - \left(\sum_{j=1}^{s} p_j \cdot u_{i,j} \right) \geq 0 \qquad (6.2.25)$$

for all i, where p_i is the estimate of the probability that the i-th alternative A_i is true. Similarly to the case of two alternatives, we conclude that

$$p_i = \frac{1}{n} \cdot \sum_{k=1}^{n} p_{i,k},$$

where n is the number of experts and $p_{i,k}$ is the estimate of the probability of the i-th alternative A_i made by the k-th expert.

When we add, to n original experts, the student as a new expert, with $p_{i,n+1} = P_i$, then the probabilities p_i change to new values

$$p_i' = \frac{1}{n+1} \cdot \sum_{k=1}^{n+1} p_{i,k} = \frac{n}{n+1} \cdot p_i + \frac{1}{n+1} \cdot P_i.$$

Thus, the left-hand side of the inequality (6.2.25) has a change which is linear in terms of the degrees $P_1, \ldots, P_s$.

For each case when the addition of the student as a new expert changes the inequality between two expected utilities, the corresponding interval of possible values of the difference is small and thus, the resulting utility is proportional to the linear function of P_i—and is, thus, linear as well.

The probability of each such change is very small, so the probability that the addition of a student can change two or more inequalities—i.e., that two changes can occur at the same time—can be estimated as the product of these two (or more) small numbers and can, therefore, be safely ignored.

In this approximation, the overall utility can be obtained by adding the probabilities of all such cases and is, therefore, also a linear function of the probabilities P_i:

$$u = u_0 + \sum_{i=1}^{s} a_i \cdot P_i.$$

Let c be the index of the correct answer, then this formula can be reformulated as

$$u = u_0 + u_c \cdot P_c + \sum_{i \neq c} a_i \cdot P_i. \qquad (6.2.26)$$

Usually, there are no a priori reasons why one incorrect answer is better than another incorrect answer. So, it is natural to assume that the utility c_i corresponding to each incorrect answer A_i is the same. Let us denote this common value of the utility by f. Then, the above formula (6.2.26) takes the form

$$u = u_0 + u_c \cdot P_c + f \cdot \sum_{i \neq c} P_i. \tag{6.2.27}$$

Since the degrees P_i add up to 1, we have

$$P_c + \sum_{i \neq c} P_i = 1,$$

hence

$$\sum_{i \neq c} P_i = 1 - P_c,$$

and the formula (6.2.27) takes the form

$$u = u_0 + u_c \cdot P_c + f \cdot (1 - P_c).$$

Thus, the utility is a linear function of the student's degree of confidence P_c in the correct answer.

Let us denote by N the number of points that we assign to a correct answer in which the student is fully confident ($P_c = 1$). Naturally, a student gets 0 points when he or she is fully confident in the wrong answer (i.e., $P_i = 1$ for some $i \neq c$ and thus, $P_c = 0$). Thus, the desired linear function should be equal to N when $P_c = 1$ and to 0 when $P_c = 0$. There is only one such linear function: $N \cdot P_c$. So, we arrive at the following recommendation.

What if there are several possible answers: the resulting recommendation. Let $A_1, \ldots, A_s$ be possible answers, out of which only one answer A_c is correct. Let N denote the number of points that a student would get for a correct answer in which he or she is absolutely confident.

During the test, the student assigns, to each possible answer A_i, his/her degree of confidence P_i that this answer is correct. These degrees must add up to 1: $\sum_{i=1}^{s} P_i = 1$.

Our analysis shows that for this, we give the student $N \cdot P_c$ points, where P_c is the student's degree of confidence in the correct answer.

How do we combine grades corresponding to different problems? In the above text, we describe how to assign number of points to a *single* question. Namely:

- Our idea was to assign the number of points which is proportional to the gain in expected utility that the student's answer can bring in a real decision making situation.

- Our analysis has shown that this expected utility is proportional to the probability P_1 of the correct answer.
- Thus, our recommendation is to assign the number of points proportional to the probability of the correct answer.

Real-life tests usually have *several* questions.

- In the traditional setting, a student provides answers to each of these questions.
- In the new setting, for each question, the student also provides us with his/her degrees of confidence in different possible answers to this question.

How do we gauge the student knowledge level based on all this information?

A natural idea is—similarly to the case of a single question—to use, as the measure of the student's knowledge, the expected utility that the student's answers can bring in a real decision making situation. Let us show how this idea can be applied.

The general decision making situation means selecting a decision for each of the problems. For example, on a medical exam, a student may be asked several questions describing different patients.

Usually, different questions on the test are independent from each other. It is known (see, e.g., [36]) that if a decision problem consists of several independent decisions, then the utility of each combination of selections is equal to the sum of the corresponding utilities.

We know the utility corresponding to each question—this is the value that we used as a recommended grade for this particular question. Thus, the overall grade for the test should be equal to the sum of the grades corresponding to individual questions.

Hence, we arrive at the following recommendation.

Resulting recommendation. Let us consider a test with T questions $q_1, \ldots, q_t, \ldots,$ q_T. For each question q_t, a student is given several possible answers $A_{t,1}, A_{t,2}, \ldots$. For each question q_t, we know the number of points N_t corresponding to the answer which is correct and for which the student has a full confidence.

The student is required, for each question q_t and for each possible answer $A_{t,i}$, to provide his/her degree of confidence $P_{t,i}$ that this particular answer is correct. For each question q_t, these probabilities should add up to 1: $P_{t,1} + P_{t,2} + \cdots = 1$.

To estimate the student's level of knowledge, we need to know, for each question q_t, the correct answer; let us denote this correct answer by $A_{t,c(t)}$. Then:

- for each question q_t, we give the student $P_{t,c(t)} \cdot N_t$ points;
- as an overall grade g, i.e., as a measure of overall student knowledge, we take the sum of the points given for individual problems: $g = \sum_{t=1}^{T} P_{t,c(t)} \cdot N_t$.

6.2.3 How to Combine Grades for Different Classes into a Single Number: Need to Go Beyond GPA

At present, the amounts of knowledge acquired by different graduates of the same program are usually compared by comparing their Grade Point Averages (GPAs). In this subsection, we argue that this is not always the most adequate description: for example, if, after completing all required classes with the highest grade of "excellent" (A), a student takes an additional challenging class and gets a "satisfactory" grade (C), the amount of his/her knowledge increases, but the GPA goes down. We propose a modification of the GPA which is free of this drawback and is, thus, more adequate for describing the student's knowledge. We also provide a psychological explanation for why people cling to the traditional GPA.

The results from this subsection first appeared in [76].

How graduate's knowledge is compared now. At present, the amounts of knowledge acquired by different graduates of the same program are usually compared by comparing their Grade Point Average (GPA). A GPA is simply an arithmetic average of all the grades that a student got in different classes—usually weighted by the number of credit hours corresponding to each class.

Usually, two types of GPA are considered:

- the overall GPA that counts all the classes that the student took at the university, and
- the major GPA, in which general education classes are not counted, the only classes which are counted as classes directly related to the student's major.

Why this is not always adequate. Let us give an example explaining why this may not be the most adequate way of comparing the students' knowledge.

Let us assume that we have two students; both took all the classes required for a Computer Science degree, and both got "excellent" grades (As) in all these classes.

The first student did not take any other classes, while the second student decided to also take an additional—very challenging—class, and got a satisfactory grade of C in this additional class.

From the common sense viewpoint, the second student knows everything that the first student knows, plus she also knows some additional material that she learned in this challenging class. However, the GPA is higher for the first student, so, from the usual GPA viewpoint, the first student is academically better.

How can we modify the GPA to make it more adequate: a proposal. To avoid the above counterintuitive situation, we propose the following natural modification of the GPA:

- if the student took only the classes needed for graduation, then his/her GPA is computed in exactly the same way as usual;
- if a student took an additional class which is not related to his/her major and is not required for his/her degree, the grade for this class should be simply ignored;

- if for some topic needed for the major, the student took more classes than required, then only the the required number of top grades are counted when computing the modified GPA.

For example, if, instead of required three technical electives, a student took four classes and got grades A, C, A, and B, then only the top three grades (two As and a B) are counted.

Why this works. If a student, in addition to all As for required classes, gets a C for an additional non-required technical elective, this C grade will not count towards a newly modified GPA.

If this is a good idea, why is it not used? As we have mentioned earlier, people often substitute average instead of the sum. From this viewpoint, when estimating the overall knowledge of a students, they tend to use the average grade (GPA) instead of, e.g., the sum of all the grades.

Just like, when pricing two large dinnerware sets,

- one consisting of 24 pieces in perfect condition, and
- the other consisting of the same 24 pieces plus 16 additional pieces, 8 of which are broken,

most people value the second set lower—although we could simply throw away the broken pieces and consider only the 24 most desirable pieces—based on the GPA, when comparing two students:

- a student who took only the required classes and got As in all of them, and
- a student who, in additional to all these classes (for which he or she also got As), also took an additional class for which this student received a B,

we select the first student as a better one.

Chapter 7
Conclusions and Future Work

While in general, humans behave rationally, there are many known experiments in which humans show seemingly irrational behavior. For example, when a customer is presented with two objects, one somewhat cheaper and another one more expensive and of higher quality, the customer's choice often depend on the presence of the third object, the object that the customer will not select:

- if the third object is cheaper than either of the two, the customer will usually select the cheaper of the two objects;
- if the third object is more expensive than the either of the two, the customer will usually select the more expensive of the two objects.

From the rational viewpoint, the selection between the two objects should not depend on the presence of other, less favorable, objects—but it does!

There are many other examples of such seemingly irrational human behavior. This phenomenon is known since the 1950s, and an explanation for this phenomenon is also well known—such seemingly irrational behavior is caused by the fact that human computational abilities are limited; in this sense, human rationality is *bounded*.

The idea of bounded rationality explains, on the *qualitative* level, why human behavior and decision making are sometimes seemingly irrational. However, until recently, there have been few successful attempts to use this idea to explain *quantitative* aspects of observed human behavior. In this book, we show, on several examples, that these quantitative aspects can be explained if we take into account that one of the main consequences of bounded rationality is *granularity*.

The main idea behind granularity is that since we cannot process *all* the available information, we only process *part* of it. Because of this, several different data points—differing by the information that we do not process—are treated the same way. In other words, instead of dealing with the original data points, we deal with *granules*, each of which corresponds to several possible data points. For example, if we only

© Springer International Publishing AG 2018
J. Lorkowski and V. Kreinovich, *Bounded Rationality in Decision Making Under Uncertainty: Towards Optimal Granularity*, Studies in Systems, Decision and Control 99, DOI 10.1007/978-3-319-62214-9_7

use the first binary digit x_1 in the binary expansion of a number $x = 0.x_1x_2\ldots$ from the interval $[0, 1]$, this means that, instead of the exact number x, we use two granules corresponding to intervals $[0, 0.5)$ (for which $x_1 = 0$) and $[0.5, 1]$ (for which $x_1 = 1$).

In this book, we have shown, on several examples (including the above customer examples) that granularity indeed explained the observed quantitative aspects of seemingly irrational human behavior. We also showed that similar arguments explain other aspects of human decision making, as well as the success of heuristic techniques in expert decision making. We then used these explanations to predict the quality of the resulting decisions. Finally, we explained how we can improve on the existing heuristic techniques by formulating and solving the corresponding optimization problems.

The main remaining challenge is to provide a similar explanation for other observed cases of seemingly irrational human behavior and decision making.

Appendix A
Computational Aspects of Decision Making Under Uncertainty

In many practical problems, we need to estimate the range of a given expression $f(x_1, \ldots, x_n)$ when each input x_i belongs to a known interval $[\underline{x}_i, \overline{x}_i]$—or when each input x_i is described by a known fuzzy set. It is known that this problem is easy to solve when we have a Single Use Expression, i.e., an expression in which each variable x_i occurs only once. In this section, we show that for similarly defined Double Use Expressions, the corresponding range estimation problem is NP-hard. Similar problems are analyzed for the problem of solving linear systems under interval (and fuzzy) uncertainty.

The results of this section first appeared in [73, 79].

A.1 Importance of Interval Computations and the Role of Single Use Expressions (SUE)

Need for data processing. In many real-life situations, we need to process data, i.e., use the estimated values $\widetilde{x}_1, \ldots, \widetilde{x}_n$ to estimate the value $\widetilde{y}$ of another quantity.

This may happen because we are interested in the value of a quantity that is difficult or even impossible to measure directly—e.g., the amount of oil in a well or the distance to a faraway star—but which can be estimated based on some related easier-to-measure quantities (e.g., the angles to the star from two different telescopes).

It can be because we are trying to predict the future values of some quantities based on the their current values and the known dynamical equations—e.g., if we want to predict tomorrow's weather based on today's meteorological measurements.

In all these cases, we apply an appropriate algorithm f to the known estimates and get the desired estimate $\widetilde{y} = f(\widetilde{x}_1, \ldots, \widetilde{x}_n)$. This algorithm can be as simple as applying an explicit formula (to find the distance to a star) or as complex as solving a system of partial differential equations (to predict weather).

© Springer International Publishing AG 2018
J. Lorkowski and V. Kreinovich, *Bounded Rationality in Decision Making Under Uncertainty: Towards Optimal Granularity*, Studies in Systems, Decision and Control 99, DOI 10.1007/978-3-319-62214-9

Need for interval data processing and taking uncertainty into account. Estimates are never absolutely accurate: for each of the input quantities the estimate $\widetilde{x}_i$ is, in general, different from its actual (unknown) value x_i. As a result, even if the algorithm f is exact—i.e., it would have produced the exact value $y = f(x_1, \ldots, x_n)$ if we plug in the exact values x_i—because of the uncertainty $\widetilde{x}_i \neq x_i$, the value $\widetilde{y}$ is, in general, different from the desired value y.

It is therefore necessary to analyze how the uncertainty in estimating x_i affects the uncertainty with which we determine y.

Need for interval data processing and interval computations. When estimates come from measurements, the difference $\Delta_i = \widetilde{x}_i - x_i$ is, is called a *measurement error*.

Sometimes, we know the probabilities of different values of measurement errors, but often, the only information that we have about the measurement error Δx_i is the upper bound Δ_i provided by the manufacturer: $|\Delta x_i| \leq \Delta_i$; see, e.g., [132]. In such situations, the only information that we have about x_i is that x_i belongs to the interval

$$\mathbf{x}_i = [\widetilde{x}_i - \Delta_i, \widetilde{x}_i + \Delta_i].$$

Different values x_i from these intervals $\mathbf{x}_i$ lead, in general, to different values $y = f(x_1, \ldots, x_n)$. So, to gauge the uncertainty in y, it is necessary to find the range of all possible values of y:

$$\mathbf{y} = [\underline{y}, \overline{y}] = \{f(x_1, \ldots, x_n) : x_1 \in \mathbf{x}_1, \ldots, x_n \in \mathbf{x}_n\}.$$

This range is usually denoted by $f(\mathbf{x}_1, \ldots, \mathbf{x}_n)$.

The problem of estimating this range based on given intervals $\mathbf{x}_i$ constitutes the main problem of *interval computations*; see e.g., [60, 108].

Need for fuzzy data processing. In many practical situations, estimates $\widetilde{x}_i$ come from experts. In this case, we do not have guaranteed upper bounds on the estimation error $\Delta x_i = \widetilde{x}_i - x_i$. Instead, we have expert estimates of their accuracy—estimates formulated in terms of words from natural language such as "approximately 0.1". One of the main ways to formalize such informal ("fuzzy") statements is to use *fuzzy logic* (see, e.g., [65, 120]), techniques specifically designed for the purpose of such formalization.

In fuzzy logic, to describe a fuzzy property $P(x)$ of real numbers (such as "approximately 0.1"), we assign, to every real number x, the degree $\mu_P(x) \in [0, 1]$ which, according to an expert, the number x satisfies this property: if the expert is absolutely sure, this degree is 1, else it takes value between 0 and 1. Once we know the experts' degrees $d_1, d_2, \ldots$, of different statements $S_1, S_2, \ldots$, we need to estimate the degree d to which a logical combination like $S_1 \vee S_2$ or $S_1 \& S_2$ hold. In other words, for each pair of values d_1 and d_2, we must select the estimate for $S_1 \vee S_2$— which will be denoted by $f_\vee(d_1, d_2)$, and also an estimate for $S_1 \& S_2$—which will be denoted by $f_\&(d_1, d_2)$, etc.

Natural requirements—e.g., that $S \& S$ means the same as S, that $S_1 \& S_2$ means the same as $S_2 \& S_1$, etc.—uniquely determine operations $f_\&(d_1, d_2) = \min(d_1, d_2)$ and $f_\vee(d_1, d_2) = \max(d_1, d_2)$ [65, 120].

A real number y is a possible value of the desired quantity if and only if there exist values $x_1, \ldots, x_n$ which are possible values of the input quantities and for which $y = f(x_1, \ldots, x_n)$:

$$y \text{ is possible } \Leftrightarrow$$

$$\exists x_1 \cdots \exists x_n ((x_1 \text{ is possible}) \& \cdots \& (x_n \text{ is possible}) \& y = f(x_1, \ldots, x_n)).$$

Once we know the degrees $\mu_i(x_i)$ corresponding to the statements "x_i is possible", we can then use the above "and"- and "or"- operations $f_\&(d_1, d_2) = \min(d_1, d_2)$ and $f_\vee(d_1, d_2) = \max(d_1, d_2)$ (and the fact that an existential quantifier $\exists$ is, in effect, an infinite "or") to estimate the degree $\mu(y)$ to which y is possible:

$$\mu(y) = \max\{\min(\mu_1(x_1), \ldots, \mu_n(x_n)) : y = f(x_1, \ldots, x_n)\}.$$

This formula was first proposed by Zadeh, the father of fuzzy logic, and is usually called *Zadeh's extension principle*.

From the computational viewpoint, fuzzy data processing can be reduced to interval data processing. An alternative way to describe a membership function $\mu_i(x_i)$ is to describe, for each possible values $\alpha \in [0, 1]$, the set of all values x_i for which the degree of possibility is at least α. This set

$$\{x_i : \mu_i(x_i) \geq \alpha\}$$

is called an *alpha-cut* and is denoted by $X_i(\alpha)$.

It is known (see, e.g., [65, 120]), that for the alpha-cuts, Zadeh's extension principle takes the following form: for every α, we have

$$R(\alpha) = \{R(x_1, \ldots, x_n) : x_i \in X_i(\alpha)\}.$$

Thus, for every α, finding the alpha-cut of the resulting membership function $\mu(R)$ is equivalent to applying interval computations to the corresponding intervals $X_1(\alpha), \ldots, X_n(\alpha)$.

Because of this reduction, in the following text, we will only consider the case of interval uncertainty.

In general, interval computations are NP-hard. In general, the main problem of interval computations is NP-hard–meaning that, if (as most computer scientists believe) $P \neq NP$, no algorithm can always compute the desired range in feasible time (i.e., in time which is bounded by a polynomial of the length of the input).

Thus, every feasible algorithm for estimating the range **y** sometimes leads to an overestimation or an underestimation.

Comment. NP-hardness of interval computations was first proven in [39, 40] by reducing, to this problem, a known NP-hard problem of propositional satisfiability (SAT) for propositional formulas in Conjunctive Normal Form (CNF): given an expression of the type

$$(v_1 \lor \neg v_2 \lor v_3) \& (v_1 \lor \neg v_4) \& \ldots,$$

check whether there exist Boolean (true-false) values v_i that make this formula true.

The disjunctions $(v_1 \lor \neg v_2 \lor v_3)$, $(v_1 \lor \neg v_4),\ldots$, are called *clauses*, and variables and their negations are called *literals*.

An overview of related NP-hardness results is given in [69]. Later papers showed that many simple interval computation problems are NP-hard: e.g., the problem of computing the range of sample variance

$$V = \frac{1}{n} \cdot \sum_{i=1}^{n}(x_i - E)^2,$$

where $E = \dfrac{1}{n} \cdot \sum_{i=1}^{n} x_i$; see, e.g., [32, 33].

Naive (straightforward) interval computations. Historically the first algorithm for estimating the range consists of the following. For each elementary arithmetic operation $\oplus$ like addition or multiplication, due to monotonicity, we can explicitly describe the corresponding range $\mathbf{x}_1 \oplus \mathbf{x}_2$:

$$[\underline{x}_1, \overline{x}_1] + [\underline{x}_2, \overline{x}_2] = [\underline{x}_1 + \underline{x}_2, \overline{x}_1 + \overline{x}_2];$$

$$[\underline{x}_1, \overline{x}_1] - [\underline{x}_2, \overline{x}_2] = [\underline{x}_1 - \overline{x}_2, \overline{x}_1 - \underline{x}_2];$$

$$[\underline{x}_1, \overline{x}_1] \cdot [\underline{x}_2, \overline{x}_2] = [\min(\underline{x}_1 \cdot \underline{x}_2, \underline{x}_1 \cdot \overline{x}_2, \overline{x}_1 \cdot \underline{x}_2, \overline{x}_1 \cdot \overline{x}_2), \max(\underline{x}_1 \cdot \underline{x}_2, \underline{x}_1 \cdot \overline{x}_2, \overline{x}_1 \cdot \underline{x}_2, \overline{x}_1 \cdot \overline{x}_2)];$$

$$\frac{1}{[\underline{x}_1, \overline{x}_2]} = \left[\frac{1}{\overline{x}_1}, \frac{1}{\underline{x}_1}\right] \text{ if } 0 \notin [\underline{x}_1, \overline{x}_1];$$

$$\frac{[\underline{x}_1, \overline{x}_1]}{[\underline{x}_2, \overline{x}_2]} = [\underline{x}_1, \overline{x}_1] \cdot \frac{1}{[\underline{x}_1, \overline{x}_2]}.$$

These formulas form *interval arithmetic*.

To estimate the range, we then *parse* the original algorithm f—i.e., represent it as a sequence of elementary arithmetic operations, and then replace each operation with numbers with the corresponding operation with intervals.

Sometimes we thus get the exact range, but sometimes, we only get an *enclosure*—i.e., an interval that contains the exact range but is different from it. For example, for a function $f(x_1) = x_1 \cdot (1 - x_1)$ on the interval $\mathbf{x}_1 = [0, 1]$, the actual range is $[0, 0.25]$, but naive interval computations return an enclosure. Specifically, the original algorithm can be described as the sequence of the following two steps:

$$r_1 = 1 - x_1; \quad y = x_1 \cdot r_1.$$

Thus, the resulting naive interval computations lead to

$$\mathbf{r}_1 = [1, 1] - [0, 1] = [0, 1];$$

$$\mathbf{y} = [0, 1] \cdot [0, 1] =$$

$$[\min(0 \cdot 0, 0 \cdot 1, 1 \cdot 0, 1 \cdot 1), \max(0 \cdot 0, 0 \cdot 1, 1 \cdot 0, 1 \cdot 1)] = [0, 1].$$

Comment. It should be mentioned there exist more sophisticated algorithms for computing the interval range, algorithms that produce much more accurate estimation for the ranges, and these algorithms form the bulk of interval computations results [60, 108].

Single Use Expressions. There is a known case when naive interval computations lead to an exact range—case of *Single Use Expressions* (SUE), i.e., expressions $f(x_1, \ldots, x_n)$ in which each variable occurs only once; see, e.g., [51, 60, 108].

For example, $x_1 \cdot x_2 + x_3$ is a SUE, while the above example $x_1 \cdot (1 - x_1)$ is not, because in this expression, the variable x_1 occurs twice.

Natural open problems. When can we reduce a function to the SUE form?

If this is not possible, what if we have double-use expressions, i.e., expressions in which each variable occurs at most twice? is it possible to always compute the range of such expressions in feasible time? If yes, then what about triple-use expressions?

These are the questions that we answer in this Appendix.

A.2 Functions Equivalent to Single Use Expressions

One of the main problems of interval computation is computing the range of a given function on a given box. In general, computing the exact range is a computationally difficult (NP-hard) problem, but there are important cases when a feasible algorithm for computing such a function is possible. One of such cases is the case of singe-use expressions (SUE), when each variable occurs only once. Because of this, practitioners often try to come up with a SUE expression for computing a given function. It is therefore important to know when such a SUE expression is possible. In this section, we consider the case of functions that can be computed by using only arithmetic operations (addition, subtraction, multiplication, and division). We show that when there exists a SUE expression for computing such a function, then this function is equal to a ratio of two multi-linear functions (although there are ratios of multi-linear functions for which no SUE expression is possible). Thus, if for a function, no SUE expression is possible, then we should not waste our efforts on finding a SUE expression for computing this function.

Definition A.2.1 Let n be an integer; we will call this integer *a number of inputs*.

- By an *arithmetic expression*, we mean a sequence of formulas of the type $s_1 := u_1 \odot_1 v_1, s_2 := u_2 \odot_2 v_2, \ldots, s_N := u_N \odot_N v_N$, where:
  - each u_i or v_i is either a rational number, or one of the inputs x_j, or one of the previous values s_k, $k < i$;
  - each $\odot_i$ is either addition +, or subtraction −, or multiplication ·, or division /.

- By the *value* of the expression for given inputs $x_1, \ldots, x_n$, we mean the value s_N that we get after we perform all N arithmetic operations $s_i := u_i \odot_i v_i$.

Definition A.2.2 An arithmetic expression is called a *single use expression* (or *SUE*, for short), if each variable x_j and each term s_k appear at most once in the right-hand side of the rules $s_i := u_i \odot_i v_i$.

Example An expression $1/(1 + x_2/x_1)$ corresponds to the following sequence of rules:

$$s_1 := x_2/x_1; \quad s_2 := 1 + s_1; \quad s_3 = 1/s_2.$$

One can see that in this case, each x_j and each s_k appears at most once in the right-hand side of the rules.

Definition A.2.3 We say that a function $f(x_1, \ldots, x_n)$ *can be computed by an arithmetic SUE expression* if there exists an arithmetic SUE expression whose value, for each tuple $(x_1, \ldots, x_n)$, is equal to $f(x_1, \ldots, x_n)$.

Example The function $f(x_1, x_2) = \dfrac{x_1}{x_1 + x_2}$ is not itself SUE, but it can be computed by the above SUE expression $1/(1 + x_2/x_1)$.

Definition A.2.4 A function $f(x_1, \ldots, x_n)$ is called *multi-linear* if it is a linear function of each variable.

Comment. For $n = 2$, a general bilinear function has the form

$$f(x_1, x_2) = a_0 + a_1 \cdot x_1 + a_2 \cdot x_2 + a_{1,2} \cdot x_1 \cdot x_2.$$

A general multi-linear function has the form $f(x_1, \ldots, x_n) = \sum\limits_{I \subseteq \{1, \ldots, n\}} a_I \cdot \prod\limits_{i \in I} x_i$.

For example, if we take $I = \{1, 2\}$, then:

- for $I = \emptyset$, we get the free term a_0;
- for $I = \{1\}$, we get the term $a_1 \cdot x_1$;
- for $I = \{2\}$, we get the term $a_2 \cdot x_2$, and
- for $I = \{1, 2\}$, we get the term $a_{1,2} \cdot x_1 \cdot x_2$.

Proposition A.2.1 *If a function can be computed by an arithmetic SUE expression, then this function is equal to a ratio of two multi-linear functions.*

Proposition A.2.2 *Not every multi-linear function can be computed by an arithmetic SUE expression.*

Comment. As we will see from the proof, this auxiliary result remains valid if, in our definition of a SUE expression, we also allow additional differential unary and binary operations—such as computing values of special functions of one or two variables—in addition to elementary arithmetic operations.

Proof of Proposition A.2.1 This result means, in effect, that for each arithmetic SUE expression, the corresponding function $f(x_1, \ldots, x_n)$ is equal to a ratio of two multi-linear functions. We will prove this result by induction: we will start with $n = 1$, and then we will use induction to prove this result for a general n.

$1°$. Let us start with the case $n = 1$. Let us prove that for arithmetic SUE expressions of one variable, in each rule $s_i := u_i \odot_i v_i$, at least one of u_i and v_i is a constant.

Indeed, it is known that an expression for s_i can be naturally represented as a tree:

- We start with s_i as a root, and add two branches leading to u_i and v_i.
- If u_i or v_i is an input, we stop branching, so the input will be a leaf of the tree.
- If u_i or v_i is an auxiliary quantity s_k, i.e., a quantity that comes from the corresponding rule $s_k := u_k \odot_k v_k$, then we add two branches leading to u_k and v_k, etc.

Since each x_j or s_i can occur only once in the right-hand side, this means that all nodes of this tree are different. In particular, this means that there is only one node x_j. This node is either in the branch u_i or in the branch v_i. In both cases, one of the terms u_i and v_i does not depend on x_j and is, thus, a constant.

Let us show, by (secondary) induction, that all arithmetic SUE expressions with one input are fractionally linear, i.e., have the form $f(x_1) = \dfrac{a \cdot x_1 + b}{c \cdot x_1 + d}$ with rational values a, b, c, and d. Indeed:

- the variable x_1 and a constant are of this form, and
- one can easily show that as a result of an arithmetic operation between a fractional-linear function $f(x_1)$ and a constant r, we also get an expression of this form, i.e., $f(x_1)+r, f(x_1)-r, r-f(x_1), r\cdot f(x_1), r/f(x_1)$, and $f(x_1)/r$ are also fractionally linear.

Comment. It is worth mentioning that, vice versa, each fractionally linear function $f(x_1) = \dfrac{a \cdot x_1 + b}{c \cdot x_1 + d}$ can be computed by an arithmetic SUE expression. Indeed, if $c = 0$, then $f(x_1)$ is a linear function $f(x_1) = \dfrac{a}{d} \cdot x_1 + \dfrac{b}{d}$, and is, thus, clearly SUE.

When $c \neq 0$, then this function can be computed by using the following SUE form: $f(x_1) = \dfrac{a}{c} + \dfrac{b - \dfrac{a \cdot d}{c}}{c \cdot x_1 + d}$.

$2°$. Let us now assume that we already proved this result for $n = k$, and we want to prove it for functions of $n = k + 1$ variables. Since this function can be computed by an arithmetic SUE expression, we can find the first stage on which the intermediate result depends on all n variables. This means that this result comes from applying an arithmetic operation to two previous results both of which depended on fewer than n variables. Each of the two previous results thus depends on $< k + 1$ variables, i.e., on $\leq k$ variables. Hence, we can conclude that each of these two previous results is a ratio of two multi-linear functions.

Since this is SUE, the two privious results depend on non-intersecting sets of variables. Without losing generality, let $x_1, \ldots, x_f$ be the variables used in the first of these previous results, and $x_{f+1}, \ldots, x_n$ are the variables used in the second of these two previous results. Then the two previous results have the form $\dfrac{N_1(x_1 \ldots, x_f)}{D_1(x_1, \ldots, x_f)}$ and $\dfrac{N_2(x_{f+1} \ldots, x_n)}{D_2(x_{f+1}, \ldots, x_n)}$, where N_i and D_i are bilinear functions. For all four arithmetic operations, we can see that the result of applying this operation is also a ratio of two multi-linear functions. For addition, we have:

$$\frac{N_1(x_1 \ldots, x_f)}{D_1(x_1, \ldots, x_f)} + \frac{N_2(x_{f+1} \ldots, x_n)}{D_2(x_{f+1}, \ldots, x_n)} =$$

$$\frac{N_1(x_1 \ldots, x_f) \cdot D_2(x_{f+1}, \ldots, x_n) + D_1(x_1 \ldots, x_f) \cdot N_2(x_{f+1}, \ldots, x_n)}{D_1(x_1 \ldots, x_f) \cdot D_2(x_{f+1}, \ldots, x_n)}.$$

For subtraction, we have:

$$\frac{N_1(x_1 \ldots, x_f)}{D_1(x_1, \ldots, x_f)} - \frac{N_2(x_{f+1} \ldots, x_n)}{D_2(x_{f+1}, \ldots, x_n)} =$$

$$\frac{N_1(x_1 \ldots, x_f) \cdot D_2(x_{f+1}, \ldots, x_n) - D_1(x_1 \ldots, x_f) \cdot N_2(x_{f+1}, \ldots, x_n)}{D_1(x_1 \ldots, x_f) \cdot D_2(x_{f+1}, \ldots, x_n)}.$$

For multiplication, we have:

$$\frac{N_1(x_1 \ldots, x_f)}{D_1(x_1, \ldots, x_f)} \cdot \frac{N_2(x_{f+1} \ldots, x_n)}{D_2(x_{f+1}, \ldots, x_n)} = \frac{N_1(x_1 \ldots, x_f) \cdot N_2(x_{f+1}, \ldots, x_n)}{D_1(x_1 \ldots, x_f) \cdot D_2(x_{f+1}, \ldots, x_n)}.$$

For division, we have:

$$\left(\frac{N_1(x_1\ldots,x_f)}{D_1(x_1,\ldots,x_f)}\right):\left(\frac{N_2(x_{f+1}\ldots,x_n)}{D_2(x_{f+1},\ldots,x_n)}\right)=\frac{N_1(x_1\ldots,x_f)\cdot D_2(x_{f+1},\ldots,x_n)}{D_1(x_1\ldots,x_f)\cdot N_2(x_{f+1},\ldots,x_n)}.$$

After that, we perform arithmetic operations between a previous result and a constant—since neither of the n variables can be used again.

Similar to Part 1 of this proof, we can show that the result of an arithmetic operation between a ratio $f(x_1, x_2, \ldots, x_n)$ of two multi-linear functions and a constant r, we also get a similar ratio.

The proposition is proven.

Proof of Proposition A.2.2 Let us prove, by contradiction, that a bilinear function $f(x_1, x_2, x_3) = x_1 \cdot x_2 + x_2 \cdot x_3 + x_2 \cdot x_3$ cannot be computed by a SUE expression. Indeed, suppose that there is a SUE expression that computes this function. By definition of SUE, this means that first, we combine the values of two of these variables, and then we combine the result of this combination with the third of the variables. Without losing generality, we can assume that first we combine x_1 and x_2, and then add x_3 to this combination, i.e., that our function has the form $f(x_1, x_2, x_3) = F(a(x_1, x_2), x_3)$ for some functions $a(x_1, x_2)$ and $F(a, x_3)$.

The function obtained on each intermediate step is a composition of elementary (arithmetic) operations. These elementary operations are differentiable, and thus, their compositions $a(x_1, x_2)$ and $F(a, x_3)$ are also differentiable. Differentiating the above expression for f in terms of F and a by x_1 and x_2, we conclude that

$$\frac{\partial f}{\partial x_1} = \frac{\partial F}{\partial a}(a(x_1, x_2), x_3) \cdot \frac{\partial a}{\partial x_1}(x_1, x_2)$$

and

$$\frac{\partial f}{\partial x_2} = \frac{\partial F}{\partial a}(a(x_1, x_2), x_3) \cdot \frac{\partial a}{\partial x_2}(x_1, x_2).$$

Dividing the first of these equalities by the second one, we see that the terms $\dfrac{\partial F}{\partial a}$ cancel each other. Thus, the ratio of the two derivatives of f is equal to the ratio of two derivatives of a and therefore, depends only on x_1 and x_2:

$$\frac{\dfrac{\partial f}{\partial x_1}}{\dfrac{\partial f}{\partial x_2}} = \frac{\dfrac{\partial a}{\partial x_1}(x_1, x_2)}{\dfrac{\partial a}{\partial x_1}(x_1, x_2)}.$$

However, for the above function $f(x_1, x_2, x_3)$, we have $\dfrac{\partial f}{\partial x_1} = x_2 + x_3$ and $\dfrac{\partial f}{\partial x_2} = x_1 + x_3$. The ratio $\dfrac{x_2 + x_3}{x_1 + x_3}$ of these derivatives clearly depends on x_3 as well—and we showed that in the SUE case, this ratio should only depend on x_1 and x_2. The contradiction proves that this function cannot be computed by a SUE expression. The proposition is proven.

A.3   From Single to Double Use Expressions

Analysis of the problem and the main result. Since the original proof of NP-hardness of interval computations comes from reduction to SAT, let us consider the corresponding SAT problems. Namely, we will say that

- a propositional formula of the above type is a Single-Use Expression (SUE) if in this formula, each Boolean variable occurs only once; and
- a Double-Use Expression (DUE) if each Boolean variable occurs at most twice.

For example:

- $(v_1 \vee \neg v_2 \vee v_3) \,\&\, (v_4 \,\&\, v_5)$ is a SUE formula, and
- $(v_1 \vee \neg v_2 \vee v_3) \,\&\, (v_1 \,\&\, \neg v_3)$ is a DUE formula: here v_1 and v_3 occur twice, and v_2 occurs once.

For propositional formulas, checking satisfiability of SUE formulas is feasible:

Proposition A.3.1 *There exists a feasible algorithm for checking propositional satisfiability of SUE formulas.*

Comment. For reader's convenience, all the proofs are placed at the end of this appendix.

For DUE formulas, we have a similar result:

Proposition A.3.2 *There exists a feasible algorithm for checking propositional satisfiability of DUE formulas.*

One may thus expect that the interval computations problem for DUE expressions is also feasible. However, our result is opposite:

Proposition A.3.3 *The main problem of interval computations for DUE formulas is NP-hard.*

Systems of interval linear equations: reminder. In many cases, instead of a known algorithm, we only have implicit relations between the inputs x_i and the desired value y. The simplest such case is when these relations are linear, i.e., when we need to determine the desired values $y_1, \ldots, y_n$ from the system of equations

$$\sum_{j=1}^{n} a_{ij} \cdot y_j = b_i,$$

where we know estimates for a_{ij} and b_i—e.g., intervals $\mathbf{a}_{ij}$ and $\mathbf{b}_i$ of possible values of these variables.

In this case, a natural question is to find the range of all possible values y_j when a_{ij} takes values from $\mathbf{a}_{ij}$ and b_i takes values from the interval $\mathbf{b}_i$.

Systems of interval linear equations: what is known about their computational complexity. It is known that computing the desired ranges is an NP-hard problem; see, e.g., [69].

However, a related problem is feasible: given a sequence of values $x_1, \ldots, x_n$ check whether there exist values $a_{ij} \in \mathbf{a}_{ij}$ and $b_i \in \mathbf{b}_i$ for which the above system is true.

This algorithm can be easily described in SUE terms: for every i, the expression $\sum_{j=1}^{n} a_{ij} \cdot y_j$ is a SUE, thus, its range can be found by using naive interval computation, as $\sum_{j=1}^{n} \mathbf{a}_{ij} \cdot y_j$. The above equality is possible if and only if this range and the interval $\mathbf{b}_i$ have a non-empty intersection for every i:

$$\left(\sum_{j=1}^{n} \mathbf{a}_{ij} \cdot y_j \right) \cap \mathbf{b}_i \neq \emptyset.$$

Checking whether two intervals $[\underline{x}_1, \overline{x}_1]$ and $[\underline{x}_2, \overline{x}_2]$ have a non-empty intersection is easy:

$$[\underline{x}_1, \overline{x}_1] \cap [\underline{x}_2, \overline{x}_2] \neq \emptyset \Leftrightarrow \underline{x}_1 \leq \overline{x}_2 \,\&\, \underline{x}_2 \leq \overline{x}_1.$$

Thus, we indeed have a feasible algorithm; this criterion is known as the Oettli-Prager criterion [60, 108].

Parametric interval linear systems: reminder. In some cases, we have additional constraints on the values a_{ij}. For example, we may know that the matrix a_{ij} is symmetric: $a_{ij} = a_{ji}$. In this case, not all possible combinations $a_{ij} \in \mathbf{a}_{ij}$ are allowed: only those for which $a_{ij} = a_{ji}$. In this case, it is sufficient to describe the values a_{ij} for $i \leq j$, the others can be expressed in terms of these ones.

In general, we can consider a *parametric* system in which we have k parameters $p_1, \ldots, p_k$ that take values from known intervals $\mathbf{p}_1, \ldots, \mathbf{p}_k$, and values a_{ij} and b_i

are linear functions of these variables: $a_{ij} = \sum\limits_{\ell=1}^{k} a_{ij\ell} \cdot p_\ell$ and $b_i = \sum\limits_{\ell=1}^{k} b_{i\ell} \cdot p_\ell$, for known coefficients $a_{ij\ell}$ and $b_{i\ell}$.

Parametric interval linear systems: what is known about their computational complexity. This problem is more general than the above problem of solving systems of linear equations. Thus, since the above problem is NP-hard, this problem is NP-hard as well.

The next natural question is: is it possible to check whether a given tuple $x = (x_1, \ldots, x_n)$ is a solution to a given parametric interval linear system, i.e., whether there exist values p_ℓ for which $\sum\limits_{j=1}^{n} a_{ij} \cdot y_j = b_i$.

The first result of this type was proven in [130, 131]. In these papers, it is shown that if each parameter p_i occurs only in one equation (even if it occurs several times in this equation), then checking is still feasible.

The proof can also be reduced to the SUE case: indeed, in this case, it is sufficient to consider one equation at a time—since no two equations share a parameter. For each i, the corresponding equation $\sum\limits_{j=1}^{n} a_{ij} \cdot y_j = b_i$ takes the form

$$\sum_{j=1}^{n} \sum_{\ell=1}^{k} a_{ij\ell} \cdot y_j \cdot p_\ell = \sum_{\ell=1}^{k} b_{i\ell} \cdot p_\ell,$$

i.e., the (SUE) linear form

$$\sum_{\ell=1}^{k} A_{i\ell} \cdot p_\ell = 0,$$

where

$$A_{i\ell} = \sum_{j=1}^{n} a_{ij\ell} \cdot y_j - b_{i\ell},$$

and we already know that checking the solvability of such an equation is feasible.

Natural questions. What happens if we we allow each parameter to occur several times? What if each parameter occurs only in one equation, but the dependence of a_{ij} and b_i on the parameters can be quadratic (this question was asked by G. Alefeld):

$$a_{ij} = a_{ij0} + \sum_{\ell=1}^{k} a_{ij\ell} \cdot p_\ell + \sum_{\ell=1}^{k} \sum_{\ell'=1}^{k} a_{ij\ell\ell'} \cdot p_\ell \cdot p_{\ell'};$$

$$b_i = b_{i0} + \sum_{\ell=1}^{k} b_{i\ell} \cdot p_\ell + \sum_{\ell=1}^{k} \sum_{\ell=1}^{k} b_{i\ell\ell'} \cdot p_\ell \cdot p_{\ell'}.$$

In this section, we provide answers to both questions.

Proposition A.3.4 *When we only allow linear dependence on parameters, then there exists a feasible algorithm that checks whether a given tuple x belongs to a solution set of a parametric interval linear system.*

Proposition A.3.5 *For parametric interval linear systems with quadratic dependence on parameters, the problem of checking whether a given tuple x belongs to a solution set of a given system is NP-hard even if we only consider systems in which each parameter occurs only on one equation.*

Proof of Proposition A.3.1 This algorithm is simple because every SUE propositional formula is satisfiable. Indeed, each variable v_i occurs only once.

- If it occurs as negation $\neg v_i$, then we can set v_i to false, after which $\neg v_i$ becomes true.
- If the variable v_i occurs without negation, then we set v_i to be true.

In both cases, for this choice, all the literals v_i or $\neg v_i$ are true, and thus, the whole formula is true.

Proof of Proposition A.3.2 Let us show that we can "eliminate" each variable v_i— i.e., in feasible time, reduce the problem of checking satisfiability of the original formula to the problem of checking satisfiability of a formula of the same (or smaller) length, but with one fewer variable.

Indeed, since the formula is DUE, each variable v_i occurs at most twice.

If it occurs only once as $\neg v_i$, then the formula has the form $(\neg v_i \vee r) \,\&\, R$, where r denotes the remaining part of the clause containing $\neg v_i$, and R is the conjunction of all the other literals. Let us show that the satisfiability of the original formula is equivalent to satisfiability of a shorter formula R that does not contain v_i at all. Indeed:

- If the original formula $(\neg v_i \vee r) \,\&\, R$ is satisfied, this means that it is true for some selection of variables. For this same selection of variables, R is true as well, so the formula R is also satisfied.
- Vice versa, let us assume that R is satisfied. This means that for some selection of variables, R is true. If we now take v_i to be false, then the clause $(\neg v_i \vee r)$ will be true as well, and thus, the whole formula $(\neg v_i \vee r) \,\&\, R$ will be true.

Similarly, if the variable v_i occurs once as v_i, then the formula has the form $(v_i \vee r) \,\&\, R$, and its satisfiability is equivalent to satisfiability of a shorter formula R that does not contain v_i at all.

If the variable v_i occurs twice, and both times as v_i, then the formula has the form $(v_i \vee r) \,\&\, (v_i \vee r') \,\&\, R$, and its satisfiability is equivalent to satisfiability of a shorter formula R that does not contain v_i at all.

If the variable v_i occurs twice, and both times as $\neg v_i$, then the formula has the form $(\neg v_i \vee r) \,\&\, (\neg v_i \vee r') \,\&\, R$, and its satisfiability is equivalent to satisfiability of a shorter formula R that does not contain v_i at all.

Finally, if it occurs once as v_i and once as $\neg v_i$, i.e., if it has the form two clauses $(v_i \vee r) \& (\neg v_i \vee r') \& R$, then its satisfiability is equivalent to the satisfiability of the new formula $(r \vee r') \& R$ (this fact is known as *resolution rule*). Indeed:

- If the formula $(r \vee r') \& R$ is satisfied, this means that for some combination of variables, both R and $r \vee r'$ are true. Thus, either r is true, or r' is true.

  - In the first case, we can take v_i to be false, then both $v_i \vee r$ and $\neg v_i \vee r'$ are true.
  - In the second case, we can take v_i to be true, then both $v_i \vee r$ and $\neg v_i \vee r'$ are true.

  Thus, in both cases, the formula $(v_i \vee r) \& (\neg v_i \vee r') \& R$ is true as well.
- Vice versa, if the original formula $(v_i \vee r) \& (\neg v_i \vee r') \& R$ is satisfied by some selection of the values, then, in this selection, either v_i is true or it is false.

  - In the first case, from the fact that $\neg v_i \vee r'$ is true and $\neg v_i$ is false, we conclude that r' is true. Thus, the disjunction $r \vee r'$ is also true.
  - In the second case, from the fact that $v_i \vee r$ is true and v_i is false, we conclude that r is true. Thus, the disjunction $r \vee r'$ is also true.

  Thus, in both cases, the formula $(r \vee r') \& R$ is satisfied as well.

The proposition is proven.

Proof of Proposition A.3.3 As we mentioned, computing the range of variance under interval uncertainty is NP-hard, but variance is a DUE:

$$V = \frac{x_1^2 + \ldots + x_i^2 + \ldots + x_n^2}{n} - \left(\frac{x_1 + \ldots + x_i + \ldots + x_n}{n} \right)^2.$$

The proposition is proven.

Proof of Proposition A.3.4 In this case, we need to check whether there are values p_ℓ that satisfy the system of linear equations $\sum_{\ell=1}^{k} A_{i\ell} \cdot p_\ell = 0$ and linear inequalities $\underline{p}_\ell \leq p_\ell \leq \overline{p}_\ell$ (that describe interval constraints on p_ℓ).

It is known that checking consistency of a given system of linear equations and inequalities is a feasible problem, a particular case of linear programming; see, e.g., [23]. Thus, any feasible algorithm for solving linear programming problem solves our problem as well. The proposition is proven.

Proof of Proposition A.3.5 We have already mentioned that finding the range of a quadratic function $f(p_1, \ldots, p_k)$ under interval uncertainty $p_\ell \in \mathbf{p}_\ell$, is NP-hard. It is also true (see, e.g., [69]) that checking, for a given value v_0, where there exists values $p_\ell \in \mathbf{p}_\ell$ for which $f(p_1, \ldots, p_k) = v_0$ is also NP-hard.

We can reduce this NP-hard problem to our problem by considering a very simple system consisting of a single equation $a_{11} \cdot y_1 = b_1$, with $y_1 = 1$, $b_1 = v_0$, and $a_{11} = f(p_1, \ldots, p_k)$. The tuple $x = (1)$ belongs to the solution set if and only if there exist values p_ℓ for which $f(p_1, \ldots, p_k) = v_0$.

The reduction is proven, so our checking problem is indeed NP-hard.

References

1. Ahsanullah, M., Nevzorov, V.B., Shakil, M.: An Introduction to Order Statistics. Atlantis Press, Paris (2013)
2. Aliev, R.A.: Fundamentals of the Fuzzy Logic-Based Generalized Theory of Decisions. Springer, Berlin (2013)
3. Alsina, C., Frank, M.J., Schweizer, B.: Associative Functions: Triangular Norms and Copulas. World Scientific, Singapore (2006)
4. Arnold, B.C., Balakrishnan, N., Nagaraja, H.N.: A First Course in Order Statistics, Society of Industrial and Applied Mathematics (SIAM). Pennsylvania, Philadelphia (2008)
5. Ajroud, A., Benferhat, S.: An approximate algorithm for min-based possibilistic networks. Int. J. Intell. Syst. **29**(7), 615–633 (2014)
6. Ayachi, R., Amor, N.B., Benferhat, S.: A generic framework for a compilation-based inference in probabilistic and possibilistic networks. Inf. Sci. **257**, 342–356 (2014)
7. Ayachi, R., Amor, N.B., Benferhat, S.: Inference using complied min-based possibilstic causal networks in the presence of interventions. Fuzzy Sets Syst. **239**, 104–136 (2014)
8. Babers, C.: Architecture Development Made Simple, Lulu.com (2006)
9. Becker, G.S.: A Treatise on the Family. Harvard University Press, Cambridge (1991)
10. Beevers, C.E., Wild, D.G., McGuire, G.R., Fiddes, D.J., Youngson, M.A.: Issues of partial credit in mathematical assessment by computer. Res. Learn. Technol. **7**(1), 26–32 (1999)
11. Beirlant, J., Goegevuer, Y., Teugels, J., Segers, J.: Statistics of Extremes: Theory and Applications. Wiley, Chichester (2004)
12. Beliakov, G., Pradera, A., Calvo, T.: Aggregation Functions: A Guide for Practitioners. Springer, New York (2007)
13. Bellman, R.E., Zadeh, L.A.: Decision making in a fuzzy environment. Manag. Sci. **17**(4), B141–B164 (1970)
14. Belohlavek, R., Trnecka, M.: Basic level of concepts in formal concept analysis. In: Proceedings of ICFCA'12. Springer, Berlin, pp. 28–44 (2012)
15. Belohlavek, R., Trnecka, M.: Basic level in formal concept analysis: interesting concepts and psychological ramifications. In: Proceedings of the Twenty-Third International Joint Conference on Artificial Intelligence IJCAI'2013, Beijing, China, pp. 1233–1239 (2013)
16. Benferhat, S., da, : Costa Pereira, C., Tettamanzi, A.G.B.: Syntactic computation of hybrid possibilistic conditioning under uncertain inputs. In: Rossi, F. (ed.) Proceedings of the Twenty-Third International Joint Conference on Artificial Intelligence IJCAI'2013, pp. 3–9. China, August, Beijing (2013)
17. Benferhat, S., et al.: An intrusion detection and alert correlation approach based on revising probabilistic classifiers using expert knowledge. Appl. Intell. **38**, 520–540 (2012)

© Springer International Publishing AG 2018

J. Lorkowski and V. Kreinovich, *Bounded Rationality in Decision Making Under Uncertainty: Towards Optimal Granularity*, Studies in Systems, Decision and Control 99, DOI 10.1007/978-3-319-62214-9

18. Bernheim, B.D., Stark, O.: Altruism within the family reconsidered: do nice guys finish last? Am. Econ. Rev. **78**(5), 1034–1045 (1988)

19. Brawley, J.V., Gao, S., Mills, D.: Associative rational functions in two variables. In: Jung-nickel, D., Niederreiter, H. (eds.) Finite Fields and Applications 2001, Proceedings of the Fifth International Conference on Finite Fields and Applications F_q5, Augsburg, Germany, 2–6 August 1999, pp. 43–56. Springer, Berlin (2001)

20. Butnariu, D., Klement, E.P.: Triangular Norm-Based Measures and Games wih Fuzzy Coali-tions. Kluwer, Dordrecht (1993)

21. Cohen, M.D., Huber, G., Keeney, R.L., Levis, A.H., Lopes, L.L., Sage, A.P., Sen, S., Whinston, A.B., Winkler, R.L., von Winterfeldt, D., Zadeh, L.A.: Research needs and the phenomena of decision making and operations. IEEE Trans. Syst. Man Cybern. **15**(6), 764–775 (1985)

22. Cole, A.J., Morrison, R.: Triplex: a system for interval arithmetic. Softw. Pract. Exp. **12**(4), 341–350 (1982)

23. Cormen, C.H., Leiserson, C.E., Rivest, R.L., Stein, C.: Introduction to Algorithms. MIT Press, Boston (2009)

24. Corter, J.E., Gluck, M.A.: Explaining basic categories: feature predictability and information. Psychol. Bull. **111**(2), 291–303 (1992)

25. David, H.A., Nagaraja, H.N.: Order Statistics. Wiley, New York (2003)

26. de Haan, L., Ferreira, A.: Extreme Value Theory: An Introduction. Springer, Berlin, Hiedel-berg, New York (2006)

27. Dubois, D., Lang, J., Prade, H.: Possibility theory: qualitative and quantitative aspects. In: Gabbay, D., Hogger, C.J., Robinson, J.A. (eds.) Handbook of Logic in Artifical Intelli-gence and Logic Programming, vol. 3, pp. 439–513. Oxford University Press, New York, Nonomonotonic Reasoning and Uncertain Reasoning (1994)

28. Dubois, D., Moral, S., Prade, H.: Belief change rules in ordinal and numerical uncertainty theories. In: Gabbay, D., Smets, P. (eds.) Handbook of Defeasible Reasioning and Uncertainty Management Systems, Quantified Representation of Uncertainty and Imprecision, vol. 1, pp. 311–392. Kluwer Academic Publ, Dordrecht (1998)

29. Dubois, D., Prade, H.: Possibility theory: qualitative and quantitative aspects. In: Gabbay, D., Smets, P. (eds.) Handbook of Defeasible Reasioning and Uncertainty Management Sys-tems, Quantified Representation of Uncertainty and Imprecision, vol. 1, pp. 169–226. Kluwer Academic Publ. Dordrecht (1998)

30. Dwork, C.: A firm foundation for private data analysis. Commun. Assoc. Comput. Mach. (ACM) **54**(1), 86–95 (2011)

31. Embrechts, P., Klüppelberg, C., Mikosch, T.: Modelling Extremal Events for Insurance and Finance. Springer, Berlin (2012)

32. Ferson, S.: Risk Assessment with Uncertainty Numbers: RiskCalc. CRC Press, Boca Raton (2002)

33. Ferson, S., Ginzburg, L., Kreinovich, V., Longpré, L., Aviles, M.: Exact bounds on finite populations of interval data. Reliab. Comput. **11**(3), 207–233 (2005)

34. Ferson, S., Kreinovich, V., Oberkampf, W., Ginzburg, L.: Experimental Uncertainty Estima-tion and Statistics for Data Having Interval Uncertainty. Sandia National Laboratories, Report SAND2007-0939, May 2007. http://www.ramas.com/intstats.pdf

35. Fishburn, P.C.: Utility Theory for Decision Making. Wiley, New York (1969)

36. Fishburn, P.C.: Nonlinear Preference and Utility Theory. John Hopkins Press, Baltimore (1988)

37. Fisher, D.H.: Knowledge acquisition via incremental conceptual clustering. Mach. Learn. **2**(2), 139–172 (1987)

38. Friedman, D.D.: Price Theory. South-Western Publishing, Cincinnati, Ohio (1986)

39. Gaganov, A.A.: Computational Complexity of the Range of the Polynomial in Several Vari-ables. Leningrad University, Math. Department, M.S. thesis (1981, in Russian)

40. A. A. Gaganov, "Computational complexity of the range of the polynomial in several vari-ables," *Cybernetics*, 1985, pp. 418–421

41. Gardeñes, E., Trepat, A., Janer, J.M.: SIGLA-PL/1: development and applications. In: Nickel, K.L.E. (ed.) Interval Mathematics 1980, pp. 301–315. Academic Press, New York (1980)
42. Garey, M.E., Johnson, D.S.: Computers and Intractability: A Guide to the Theory of NP-Completeness. Freeman, San Francisco (1979)
43. Gebremedhin, A.H., Manne, F., Pothen, A.: What color is your Jacobian? Graph coloring for computing derivatives. SIAM Rev. **47**(4), 629–705 (2005)
44. Gelfand, I.M., Shilov, G.E., Vilenkin, N.Y.: Generalized Functions. Academic Press, New York (1964)
45. Gosselin, F., Schyns, P.G.: Why do we SLIP to the basic level? Computational constraints and their implementation. Psychol. Rev. **108**(4), 735–758 (2001)
46. Gray, R.M.: Entropy and Information Theory. Springer, Berlin (2011)
47. Griewank, A., Walther, A.: Evaluating Derivatives: Principles and Techniques of Algorithmic Differentiation. SIMA, Philadelphia (2008)
48. Guillemin, V.M., Sternberg, S.: An algebraic model of transitive differential geometry. Bull. Am. Math. Soc. **70**(1), 16–47 (1964)
49. Gumbel, E.J.: Statistics of Extremes. Dover Publications, New York (2004)
50. Gutierrez, L., Ceberio, M., Kreinovich, V., Gruver, R.L., Peña, M., Rister, M.J., Saldaña, A., Vasquez, J., Ybarra, J., Benferhat, S.: From interval-valued probabilities to interval-valued possibilities: case studies of interval computation under constraints. In: Proceedings of the 6th International Workshop on Reliable Engineering Computing REC'2014, Chicago, Illinois, 25–28 May 2014, pp. 77–95 (2014)
51. Hansen, E.: Sharpness in interval computations. Reliab. Comput. **3**, 7–29 (1997)
52. Hansen, E., Walster, G.W.: Global Optimization Using Interval Analysis. Marcel Decker, New York (2004)
53. Hennessy, J.L., Patterson, D.A.: Computer Architecture: A Quantitative Approach. Morgan Kaufmann, Waltham (2012)
54. Holmes, L.E., Smith, L.J.: Student evaluations of faculty grading methods. J. Educ. Bus. **78**(6), 318–323 (2003)
55. Hori, H., Kanaya, S.: Utility functionals with nonpaternalistic intergerenational altruism. J. Econ. Theory **49**, 241–265 (1989)
56. Hurwicz, L.: Optimality Criteria for Decision Making Under Ignorance. Cowles Commission Discussion Paper, Statistics, No. 370 (1951)
57. International Council on Systems Engineering (INCOSE): Systems Engineering Handbook. Wiley, Hoboken (2015)
58. Isaacson, R.M., Was, C.A.: Building a metacognitive curriculum: an educational psychology to teach metacognition. Natl. Teach. Learn. Forum **19**(5), 1–4 (2010)
59. Jaynes, E.T.: Probability Theory: The Logic of Science. Cambridge University Press, Cambridge (2003)
60. Jaulin, L., Kieffer, M., Didrit, O., Walter, E.: Applied Interval Analysis. Springer, London (2001)
61. Kahneman, D.: Thinking Fast and Slow. Farrar, Straus and Giroux, New York (2011)
62. Kaucher, E.: Über Eigenschaften und Anwendungsmöglichkeiten der Erweiterten Intervallrechnung und des Hyperbolische Fastköpers über R. Computing, Supplement **1**, 81–94 (1977)
63. Keeney, R.L., Raiffa, H.: Decisions with Multiple Objectives. Wiley, New York (1976)
64. Klement, E.P., Mesiar, R., Pap, E.: Triangular Norms. Kluwer, Dordrecht (2000)
65. Klir, G., Yuan, B.: Fuzzy Sets and Fuzzy Logic. Prentice Hall, Upper Saddle River (1995)
66. Kosfeld, M., Heinrichs, M., Zak, P.J., Fischbacher, U., Fehr, E.: Oxytocin increases trust in humans. Nature **435**(2), 673–676 (2005)
67. Koshelev, M.: Trustees' and investors' behavior in the first two rounds of a trust game: overview and a (partial) explanation based on cooperative game theory. J. Adv. Comput. Intell. Intell. Inf. (JACIII) **15**(4), 438–448 (2011)
68. Kreinovich, V.: Decision making under interval uncertainty (and beyond). In: Guo, P., Pedrycz, W. (eds.) Human-Centric Decision-Making Models for Social Sciences, pp. 163–193. Springer, Berlin (2014)

69. Kreinovich, V., Lakeyev, A., Rohn, J., Kahl, P.: Computational Complexity and Feasibility of Data Processing and Interval Computations. Kluwer, Dordrecht (1998)
70. Laplante, P.A.: Requirements Engineering for Software and Systems. CRC Press, Boca Raton (2014)
71. Lee, J.J., Lukachko, S.P., Waitz, I.A., Schafer, A.: Historical and future trends in aircraft performance, cost and emissions. Annu. Rev. Energy Environ. **26**, 167–200 (2001)
72. Likert, R.: A Technique for the measurement of attitudes. Ann. Psychol. **140**, 1–55 (1932)
73. Lorkowski, J.: From single to double use expressions, with applications to parametric interval linear systems: on computational complexity of fuzzy and interval computations. In: Proceedings of the 30th Annual Conference of the North American Fuzzy Information Processing Society NAFIPS'2011, El Paso, Texas, 18–20 March 2011
74. Lorkowski, J., Aliev, R., Kreinovich, V.: Towards decision making under interval, set-valued, fuzzy, and z-number uncertainty: a fair price approach. In: Proceedings of the IEEE World Congress on Computational Intelligence WCCI'2014, Beijing, China, 6–11 July 2014
75. Lorkowski, J., Kosheleva, O., Longpre, L., Kreinovich, V.: When can we reduce multi-variable range estimation problem to two fewer-variable problems? Reliab. Comput. **21**, 1–10 (2015)
76. Lorkowski, J., Kosheleva, O., Kreinovich, V.: How to modify a Grade Point Average (GPA) to make it more adequate. Int. Math. Forum **9**(28), 1363–1367 (2014)
77. Lorkowski, J., Kosheleva, O., Kreinovich, V.: How success in a task depends on the skills level: two uncertainty-based justifications of a semi-heuristic Rasch model. In: Proceedings of the World Congress of the International Fuzzy Systems Association IFSA'2015, Joint with the Annual Conference of the European Society for Fuzzy Logic and Technology EUSFLAT'2015, Gijon, Asturias, Spain, 30 June–3 July 2015, pp. 506–511 (2015)
78. Lorkowski, J., Kosheleva, O., Kreinovich, V.: How to take into account a student's degree of certainty when evaluating the test results. In: Proceedings of the 45th ASEE/IEEE Frontiers in Education Conference FIE'2015, El Paso, Texas, 21–24 October 2015, pp. 1568–1572 (2015)
79. Lorkowski, J., Kosheleva, O., Kreinovich, V.: Every function computable by a single use expression is a ratio of two multi-linear functions. J. Uncertain Syst. **10**(1), 48–52 (2016)
80. Lorkowski, J., Kosheleva, O., Kreinovich, V., Soloviev, S.: How design quality improves with increasing computational abilities: general formulas and case study of aircraft fuel efficiency. In: Proceedings of the International Symposium on Management Engineering ISME'2014, Kitakyushu, Japan, 27–30 July 2014, pp. 33–35 (2014)
81. Lorkowski, J., Kosheleva, O., Kreinovich, V., Soloviev, S.: How design quality improves with increasing computational abilities: general formulas and case study of aircraft fuel efficiency. J. Adv. Comput. Intell. Intell. Inf. (JACIII) **19**(5), 581–584 (2015)
82. Lorkowski, J., Kreinovich, V.: Fuzzy logic ideas can help in explaining Kahneman and Tversky's empirical decision weights. In: Proceedings of the 4th World Conference on Soft Computing, Berkeley, California, 25–27 May 2014, pp. 285–289 (2014)
83. Lorkowski, J., Kreinovich, V.: How much for an interval? A set? A twin set? A p-box? A Kaucher interval? An economics-motivated approach to decision making under uncertainty. In: Abstracts of the 16th GAMM-IMACS International Symposium on Scientific Computing, Computer Arithmetic and Validated Numerics SCAN'2014, Wuerzburg, Germany, 21–26 September 2014, pp. 98–99 (2014)
84. Lorkowski, J., Kreinovich, V.: How Much for an Interval? A Set? A Twin Set? A P-Box? A Kaucher Interval? Towards an Economics-Motivated Approach to Decision Making Under Uncertainty. In: Proceedings of the 16th GAMM-IMACS International Symposium on Scientific Computing, Computer Arithmetic, and Verified Numerical Computation SCAN'2014, Wuerzburg, Germany, 21–26 September , pp. 66–76 (2014)
85. Lorkowski, J., Kreinovich, V.: Interval and symmetry approaches to uncertainty–pioneered by Wiener–help explain seemingly irrational human behavior: a case study. In: Proceedings of the 2014 Annual Conference of the North American Fuzzy Information Processing Society NAFIPS'2014, Boston, Massachusetts, 24–26 June 2014
86. Lorkowski, J., Kreinovich, V.: Likert-scale fuzzy uncertainty from a traditional decision making viewpoint: how symmetry helps explain human decision making (including seemingly irrational behavior). Appl. Comput. Math. **13**(3), 275–298 (2014)

87. Lorkowski, J., Kreinovich, V.: Likert-scale fuzzy uncertainty from a traditional decision making viewpoint: it incorporates both subjective probabilities and utility information. In: Proceedings of the Joint World Congress of the International Fuzzy Systems Association and Annual Conference of the North American Fuzzy Information Processing Society IFSA/NAFIPS'2013, Edmonton, Canada, 24–28 June 2013, pp. 525–530 (2013)

88. Lorkowski, J., Kreinovich, V.: Fuzzy logic ideas can help in explaining Kahneman and Tversky's empirical decision weights. In: Proceedings of the 4th World Conference on Soft Computing, Berkeley, California, 25–27 May 2014, pp. 285–289 (2014)

89. Lorkowski, J., Kreinovich, V.: How to gauge unknown unknowns: a possible theoretical explanation of the usual safety factor of 2. Math. Struct. Model. **32**, 49–52 (2014)

90. Lorkowski, J., Kreinovich, V.: Granularity helps explain seemingly irrational features of human decision making. In: Pedrycz, W., Chen, S.-M. (eds.) Granular Computing and Decision-Making: Interactive and Iterative Approaches, pp. 1–31. Springer, Cham, Switzerland (2015)

91. Lorkowski, J., Kreinovich, V.: Why awe makes people more generous: utility theory can explain recent experiments. J. Uncertain Syst. **10**(1), 53–56 (2016)

92. Lorkowski, J., Kreinovich, V., Aliev, R.: Towards decision making under interval, set-valued, fuzzy, and Z-number uncertainty: a fair price approach. In: Proceedings of the IEEE World Congress on Computational Intelligence WCCI'2014, Beijing, China, 6–11 July 2014

93. Lorkowski, J., Kreinovich, V., Kosheleva, O.: In: Proceedings of the IEEE Symposium Series on Computational Intelligence, Cape Town, South Africa, 7–10 December, pp. 1621–1626 (2015)

94. Lorkowski, J., Longpre, L., Kosheleva, O., Benferhat, S.: Coming up with a good question is not easy: a proof. In: Proceedings of the Annual Conference of the North American Fuzzy Information Processing Society NAFIPS'2015 and 5th World Conference on Soft Computing, Redmond, Washington, 17–19 August 2015

95. Lorkowski, J., Trnecka, M.: Similarity approach to defining basic level of concepts explained from the utility viewpoint. In: Proceedings of the Sixth International Workshop on Constraints Programming and Decision Making CoProd'2013, El Paso, Texas, 1 November 2013, pp. 17–21 (2013)

96. Lovette, M., Meyer, O., Thille, C.: The open leaning initiative: measuring the effectiveness of the OLI statistics course in accelerating student learning. J. Interact. Media Educ. **1**, 1–16 (2008)

97. Luce, R.D., Raiffa, R.: Games and Decisions: Introduction and Critical Survey. Dover, New York (1989)

98. March, J.: Bounded rationality, ambiguity, and the engineering of choice. Bell J. Econ. **9**(2), 587–608 (1978)

99. Maslova, N.B.: Private communication

100. Madsen, D.A., Madsen, D.P.: Engineering Drawing and Design. Delmar, Cengage Learning, Clifton Park (2012)

101. Masters, G.N.: The analysis of partial credit scoring. Appl. Meas. Educ. **1**(4), 279–297 (1988)

102. McKee, J., Lorkowski, J., Ngamsantivong, T.: Note on fair price under interval uncertainty. J. Uncertain Syst. **8**(3), 186–189 (2014)

103. Mendel, J.M.: Uncertain Rule-Based Fuzzy Logic Systems: Introduction and New Directions. Prentice-Hall, Upper Saddle River (2001)

104. Mendel, J.M., Wu, D.: Perceptual Computing: Aiding People in Making Subjective Judgments. IEEE Press and Wiley, New York (2010)

105. Miller, M.D.: Minds Online: Teaching Effectively with Technology. Harvard University Press, Cambridge (2014)

106. Moore, G.E.: Cramming more components onto integrated circuits. Electronics, 114–117 (1965)

107. Moore, R.E.: Interval Analysis. Prentice-Hall, Englewood Cliffs (1966)

108. Moore, R.E., Kearfott, R.B., Cloud, M.J.: Introduction to Interval Analysis. SIAM Press, Philadelphia (2009)

109. Murphy, G.L.: The Big Book of Concepts. MIT Press, Cambridge (2002)

110. Myerson, R.B.: Game Theory: Analysis of Conflict. Harvard University Press, Cambridge (1991)
111. Narukawa, Y., Torra, V.: On distorted probabilities and M-separable fuzzy measures. Int. J. Approx. Reason. **52**, 1325–1336 (2011)
112. Nash, J.: The bargaining problem. Econometrica **18**(2), 155–162 (1950)
113. Nesterov, V.M.: Interval and twin arithmetics. Reliab. Comput. **3**(4), 369–380 (1997)
114. Nguyen, H.T., Kosheleva, O., Kreinovich, V.: Decision making beyond arrow's impossibility theorem, with the analysis of effects of collusion and mutual attraction. Int. J. Intell. Syst. **24**(1), 27–47 (2009)
115. Nguyen, H.T., Kreinovich, V.: Applications of Continuous Mathematics to Computer Science. Kluwer, Dordrecht (1997)
116. Nguyen, H.T., Kreinovich, V., Lea, B.: How to combine probabilistic and fuzzy uncertainties in fuzzy control. In: Proceedings of the Second International Workshop on Industrial Applications of Fuzzy Control and Intelligent Systems, College Station, 2–4 December 1992, pp. 117–121 (1992)
117. Nguyen, H.T., Kreinovich, V., Lorkowski, J., Abu, S.: Why Sugeno lambda-measures. In: Proceedings of the IEEE International Conference on Fuzzy Systems FUZZ-IEEE'2015, Istanbul, Turkey, 1–5 August 2015
118. Nguyen, H.T., Kreinovich, V., Wu, B., Xiang, G.: Computing Statistics Under Interval and Fuzzy Uncertainty. Springer, Berlin (2012)
119. Nguyen, H.T., Kreinovich, V., Zuo, Q.: Interval-valued degrees of belief: applications of interval computations to expert systems and intelligent control. Int. J. Uncertainty Fuzziness Knowl. Based Syst. (IJUFKS) **5**(3), 317–358 (1997)
120. Nguyen, H.T., Walker, E.A.: A First Course in Fuzzy Logic. Chapman and Hall/CRC, Boca Raton (2006)
121. Nilson, L.B.: Creating Self-Regulated Learners: Strategies to Strengthen Students' Self-Awareness and Learning Skills. Stylus Publishing, Sterling (2013)
122. Open Learning Initiative (OLI): Cargenie Mellon Unuversity. http://oli.cmu.edu
123. Pal, N.R., Bezdek, J.C.: Measuring fuzzy uncertainty. IEEE Trans. Fuzzy Syst. **2**(2), 107–118 (1994)
124. Papadimitriou, C.H.: Computational Complexity. Addison Wesley, San Diego (1994)
125. Pardalos, P.M.: Complexity in Numerical Optimization. World Scientific, Singapore (1993)
126. Pedrycz, W., Skowron, A., Kreinovich, V. (eds.): Handbook on Granular Computing. Wiley, Chichester (2008)
127. Peeters, P.M., Middel, J., Hoolhorst, A.: Fuel Efficiency of Commercial Aircraft: An Overview of Historical and Future Trends. Netherlands National Aerospace Laboratory NLR, Technical report NLR-CR-2005-669 (2005)
128. Penner, J.E., Lister, D.H., Griggs, D.J., Dokken, D.J., McFarland, M.: Aviation and the Global Atmosphere: A Special Report of Intergovernmental Panel on Climate Change (IPCC) Working Groups I and III. Cambridge University Press, Cambridge (1999)
129. Piff, P., Dietze, P., Feinberg, M., Stancato, D., Keltner, D.: Awe, the small self, and prosocial behavior. J. Personal. Soc. Psychol. **108**(6), 883–899 (2015)
130. Popova, E.D.: Explicit characterization of a class of parametric solution sets. Compt. Rend. Acad. Bulg. Sci. **62**(10), 1207–1216 (2009)
131. Popova, E.D.: Characterization of parametric solution sets. In: Abstracts of the 14th GAMM-IMACS International Symposium on Scientific Computing, Computer Arithmetic, and Validated Numerics SCAN'2010, Lyon, France, 27–30 September 2010, pp. 118–119 (2010)
132. Rabinovich, S.G.: Measurement Errors and Uncertainty: Theory and Practice. Springer, Berlin (2005)
133. Raiffa, H.: Decision Analysis. Addison-Wesley, Reading (1970)
134. Raiffa, H.: Decision Analysis. McGraw-Hill, Columbus (1997)
135. Rankine, W.J.M.: A Manual on Applied Mechanics. Richard Griffin and Company, Glasgow, Scotland, p. 274 (1858). https://archive.org/stream/manualappmecha00rankrich#page/n8/mode/1up

136. Rapoport, A.: Some game theoretic aspects of parasitism and smbiosis. Bull. Math. Biophys. **18** (1956)
137. Rapoport, A.: Strategy and Conscience. Shocken Books, New York (1969)
138. Rasch, G.: Probabilistic Models for Some Intelligence and Attainment Tests, Danish Institute for Educational Research, Copenhagen, 1960, expanded edn. The University of Chicago Press, Chicago (1960)
139. Redelmeier, D., Shafir, E.: Medical decision making in situations that offer multiple alternatives. J. Am. Med. Assoc. **273**(4), 302–305 (1995)
140. Rosch, E.: Principles of categorization. In: Rosch, E., Lloyd, B.B. (eds.) Cognition and Categorization, pp. 27–48. Lawrence Erlbaum Associates, Hillsdale (1978)
141. Sabbagh, K.: Twenty-First-Century Jet: The Making and Marketing of the Boeing 777. Scribner, New York (1996)
142. Sainz, M.A., Armengol, J., Calm, R., Herrero, P., Jorba, L., Vehi, J.: Modal Interval Analysis. Springer, Berlin (2014)
143. Shafir, E., Simonson, I., Tversky, A.: Reason-based choice. Cognition **49**, 11–36 (1993)
144. Sheskin, D.J.: Handbook of Parametric and Nonparametric Statistical Procedures. Chapman & Hall/CRC, Boca Raton (2011)
145. Simon, H.A.: A behavorial model of rational choice. Q. J. Econ. **69**(1), 99–118 (1955)
146. Simon, H.A.: Rational choice and the structure of the environment. Psychol. Rev. **63**(2), 129–138 (1956)
147. Simon, H.A.: Nobel Memorial Lecture, 8 December 1978. http://www.nobelprize.org/nobel_prizes/economic-sciences/laureates/1978/simon-lecture.pdf
148. Simonson, I., Rosen, E.: Absolute Value: What Really Influences Customers in the Age of (Nearly) Perfect Information. Harper Business, New York (2014)
149. Singer, I.M., Sternberg, S.: The infinite groups of Lie and Cartan, Part I. Journal d'Analyse Mathematique **15**, 1–113 (1965). Springer
150. Smith, R.A., Hillmansen, S.: A brief historical overview of the fatigue of railway axles. Proc. Inst. Mech. Eng. **218**, 267–277 (2004)
151. Sugeno, M.: Theory of Fuzzy Integrals and Its Applications, Ph.D. Dissertation, Tokyo. Institute of Technology (1974)
152. Suppes, P., Krantz, D.M., Luce, R.D., Tversky, A.: Foundations of Measurement: Geometrical, Threshold, and Probabilistic Representations, vol. II. Academic Press, San Diego (1989)
153. Tahani, H., Keller, J.: Information fusion in computer vision using the fuzzy integral. IEEE Trans. Syst. Man Cybern. **20**(3), 733–741 (1990)
154. Tipler, F.J.: The Physics of Immortality. Doubleday, New York (1994)
155. Tversky, A., Simonson, I.: Context-dependent preferences. Manag. Sci. **39**(10), 1179–1189 (1993)
156. Ullman, D.: Mechanical Design Process, pp. 314–317. McGraw-Hill, New York (2009)
157. Walster, G.W.: Moore's Single-Use-Expression Theorem on extended real intervals. In: Abstracts of the 2002 SIAM Workshop on Validated Computing, Toronto, Canada, 23–25 May 2002
158. Wang, Z., Klir, G.J.: Fuzzy Measure Theory. Plenum Press, New York (2010)
159. Wiener, N.: A contribution to the theory of relative position. Proc. Camb. Philos. Soc. **17**, 441–449 (1914)
160. Wiener, N.: A new theory of measurement: a study in the logic of mathematics. Proc. Lond. Math. Soc. **19**, 181–205 (1921)
161. Wiener, N.: Cybernetics, or Control and Communication in the Animal and the Machine. MIT Press, Cambridge (1962)
162. Xiang, G., Ceberio, M., Kreinovich, V.: Computing population variance and entropy under interval uncertainty: linear-time algorithms. Reliab. Comput. **13**(6), 467–488 (2007)
163. Zadeh, L.A.: Fuzzy sets. Inf. Control **8**, 338–353 (1965)
164. Zadeh, L.A.: Outline of a new approach to the analysis of complex systems and decision processes. IEEE Trans. Syst. Man Cybern. **3**(1), 28–44 (1973)

165. Zadeh, L.A.: Precisiated natural language-toward a radical enlargement of the role of natural languages in information processing, decision and control. In: Wang, L., Halgamuge, S.K., Yao, X. (eds.) Proceedings of the 1st International Conference on Fuzzy Systems and Knowledge Discovery FSDK'02: Computational Intelligence for the E-Age, Singapore, 18–22 November 2002, Vol. 1, pp. 1–3 (2002)

166. Zadeh, L.A.: A new direction in decision analysis-perception-based decisions. In: Ralescu, A.L. (ed.) Proceedings of the Fourteenth Midwest Artificial Intelligence and Cognitive Sciences Conference MAICS'2003, Cincinnati, Ohio, 12–13 April 2003, pp. 1–2 (2003)

167. Zadeh, L.A.: Computing with words and perceptions—a paradigm shift in computing and decision analysis and machine intelligence. In: Wani, A., Cios, K.J., Hafeez, K. (eds.) Proceedings of the 2003 International Conference on Machine Learning and Applications ICMLA'2003, Los Angeles, California, 23–24 June 2003, pp. 3–5 (2003)

168. Zadeh, L.A.: A note on Z-numbers. Inf. Sci. **181**, 2923–2932 (2011)

169. Zeigenfuse, M.D., Lee, M.D.: A comparison of three measures of the association between a feature and a concept. In: Carlson, L., Holscher, C., Shipley, T.F. (eds.) Proceedings of the 33rd Annual Conference of the Cognitive Science Society, pp. 243–248. Austin, Texas (2011)

Index

A

Activation function, 14
Additive measure, 27
Additivity, 27, 92
Airplane design, 83
"and"-operation, 14, 122, 139
Associativity, 35
Average instead of sum, 7
    explained, 22
Awe, 41
    increases generosity, 41

B

Basic level concept, 69
Biased probability estimate, 6
    explained, 16
Bounded rationality, 3, 9, 135
Buran, 40

C

Central Limit Theorem, 16, 84
Clause, 140
Compromise effect, 4
    definition, 5
    explanation, 13
    is irrational, 5
    used to manipulate customers, 5
Computer-aided design, 83
Concept
    basic level, 69
Concept analysis, 69
Conditional expected value, 21
Confidence interval, 16
Crisp Z-number, 100

D

Decision making, 1, 89
    as optimization, 2, 11
    based on utilities, 10
    human, 2
    under uncertainty, 90
Decision theory
    traditional, 9
Degree of confidence, 27
Design quality, 83
Designing a good question
    NP-hard, 50
Distribution
    Gumbel, 85
    normal, 16, 85
Disutility, 72, 76
Double use expression, 146

E

Education, 45, 114
    asking a good question, 46
    GPA, 133
    grade point average, 133
    partial credit, 114
    Rasch model, 57
    student's degree of certainty, 125
Embedded set, 95
Embedded-set uncertainty, 95
Empathy, 43
Enclosure, 141
Entropy, 47
Estimating probability
    biased, 6
    need for, 6
Euphoria, 42
Expected utility, 90, 126

© Springer International Publishing AG 2018
J. Lorkowski and V. Kreinovich, *Bounded Rationality in Decision Making
Under Uncertainty: Towards Optimal Granularity*, Studies in Systems,
Decision and Control 99, DOI 10.1007/978-3-319-62214-9

Printed in the United States
By Bookmasters